Eric and Barbara Graham
Candiehead House
Candie
Near Avonbridge
Stirlingshire
FK1 2LD
Telephone: 0324 86243

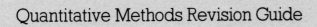

Quantitative Methods Revision Guide

Other books in this series

Cost Accounting Revision Guide   Colin Drury
Economics Revision Guide   Keith West and Rob Dixon

Quantitative Methods Revision Guide

Paul Goodwin

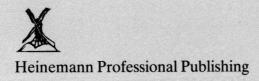

Heinemann Professional Publishing

Heinemann Professional Publishing Ltd
22 Bedford Square, London WC1B 3HH

LONDON   MELBOURNE   AUCKLAND

First published 1988

British Library Cataloguing in Publication Data
Goodwin, Paul
  Quantitative methods revision guide.
  1. Accounting – Mathematics
  I. Title
  657   RF5657

ISBN 0 434 90675 1

Typeset by Keyset Composition, Colchester
Printed and bound by Richard Clay Ltd, Chichester

Contents

Preface

This book is designed to help all accountancy students to revise effectively for examinations in Quantitative Methods. It contains a concise set of notes which enable the reader quickly to revise the essential information relating to each topic. Informal notes supplement the main text and highlight possible pitfalls and common mistakes. In addition, there is a set of past examination questions, a set of exercises with fully worked solutions, advice on examination technique and a test paper.

The material in the book reflects the latest syllabuses of the professional bodies. It is therefore ideally suited for students taking examinations for the Chartered Institute of Management Accountants, the Institutes of Chartered Accountants of England and Wales, Ireland and Scotland, the Chartered Association of Certified Accountants, the Chartered Institute of Public Finance and Accountancy and the Institute of Chartered Secretaries and Administrators. Students on Accountancy Foundation courses will also find the book useful, as will students taking degrees in Business Studies and Accountancy and Finance.

Where the syllabuses of the professional bodies differ, this is clearly indicated in the text. However, the table in Section 1 gives an overview of how the book relates to the various syllabuses.

Acknowledgements

I would like to thank the following accountancy bodies for granting permission to use tables and past examination questions.

The Chartered Institute of Management Accountants (past examination questions and the table in Appendix 1)
The Chartered Association of Certified Accountants (past examination questions)
The Institute of Chartered Accountants in England and Wales (past examination questions)

Paul Goodwin

Part One
How to Use This Book

1   The notes in Part 2 will give you the key information on each topic. The chart on page 3 shows how the topics relate to the various syllabuses, and these differences are also pointed out in the text.

   Pay special attention to the informal notes which are set next to the main text. These highlight common errors and misunderstandings.

2   Part 2 also contains a selection of past examination questions. It is advisable to attempt these without referring to the notes. If your answers disagree with those given, then re-read the appropriate section in the notes.

3   Part 3 contains a set of exercises which are designed to give you more practice in answering examination-type questions. Again, you should attempt these without referring to the notes. A fully worked solution is given to each of these questions to enable you to trace the source of any errors.

4   Part 4 explains how you should prepare for the examination and also advises on examination technique.

5   Towards the end of your revision you should take the Test Paper in Part 5. This will give you practice in answering questions under examination conditions.

Differences between syllabuses

The following abbreviations are used to indicate places in the text where syllabuses differ:

The Chartered Institute of Management Accountants:
   Stage 1, Quantitative Methods   CIMA

The Institute of Chartered Accountants of England and Wales:
  Foundation stage, Quantitative Methods   ICAEW
The Institute of Chartered Accountants of Scotland:
  Preliminary examination: Mathematical Techniques   ICAS
The Institute of Chartered Accountants in Ireland:
  Professional One, Paper 3, Statistics   ICAI
The Chartered Association of Certified Accountants:
  Syllabus 1.5, Business Mathematics and
  Information Technology   CACA(1.5)
  (This book does not cover information technology aspects of
  this syllabus)
  Syllabus 2.6, Quantitative Analysis   CACA(2.6)
The Chartered Institute of Public Finance and Accountancy:
  Professional examination 2, Analytical Techniques   CIPFA
The Institute of Chartered Secretaries and Administrators:
  Syllabus 03, Quantitative Studies   ICSA

While every effort has been made to ensure that information on the relevance of topics to syllabuses is accurate and up to date, you should of course carefully check for yourself the contents of the syllabus you are studying. Some syllabuses are phrased in rather general terms and examiners, and their interpretations of syllabuses, are likely to change over time.

The relevance of topics to the different syllabuses

Contents	CIMA	ICAEW	ICAI	ICAS	CACA(1.5)	CACA(2.6)	CIPFA	ICSA
Chapter 1								
Collecting data	*	*	*	*	*	*		*
Presenting data	*	*	*	*	*			*
Chapter 2								
Measures of location and dispersion	*	*	*	*	*			*
Chapter 3								
Time series	*	*	*	*		*	*	*
Index numbers	*	*	*		*			*
Chapter 4								
Probability theory	*	*	*	*	*	*	*	*
and decision analysis	*	*		*		*	*	
Chapter 5								
Probability distributions	*	*	*	*		*	*	*
Chapter 6								
Estimation	*	*	*	*		*	*	*
Tests based on normal distribution		*	*	*		*	*	*
t tests						*	*	
Chi-squared tests			*			*	*	
Chapter 7								
Functions and graphs	*	*			*			*
Chapter 8								
Matrix algebra	*				*			
Financial maths	*	*		*	*			*
Chapter 9								
Correlation and regression	*	*	*	*		*	*	*
Advanced topics						*		
Chapter 10								
Linear programming	*	*				*	*	*
Transportation LP						*	*	
Assignment method						*	*	
Chapter 11								
Stock control models	*			*	*	*	*	
Queuing theory						*	*	
Simulation						*	*	
Chapter 12								
CPA and PERT	*	*		*		*	*	

Part Two
Revision Notes

1
Collecting and presenting data

The material in this chapter is relevant to the following syllabuses: CIMA, ICAEW, ICAS, ICAI, CACA(1.5), CACA(2.6), ICSA

1.1 Introduction

Statistics is concerned with methods for collecting, analysing, presenting and interpreting quantitative data. A major role of statistics is the provision of information for decision making.

This chapter first considers methods of collecting data where the need to obtain reliable data has to be balanced against the cost of collecting it. Methods of presenting data are dealt with in the second part of the chapter.

1.2 Collecting data

1.2.1 Population and samples

A population is the complete collection of people, or objects (e.g. sales files, TV sets produced), which is relevant to the aims of a survey. A sample is a sub-group selected from the population. The advantages of sampling are

1 It is cheaper and quicker than a complete survey of the population
2 Since a sample is a smaller undertaking than a complete survey, it is easier to control and may involve less human error
3 When testing is destructive, as in some forms of quality control, there is no alternative to sampling.

1.2.2 Sampling frames

A sampling frame is, ideally, a complete list or collection of the members of the population (e.g. a Register of Electors, a list of members of an accountancy body or a set of cards in a filing system) from which the sample can be selected. In practice, many sampling frames are out of date and incomplete or they may include certain individuals' names more than once.

1.2.3 Probability and non-probability sampling

In probability sampling every member of the population has a chance of being selected for the sample *and* the probability of any individual being selected can be calculated. This means that probability theory can be used to assess the likely accuracy of results derived from the sample.

Probability samples require a *sampling frame*. Simple random, stratified, cluster and multi-stage samples are all forms of probability sampling.

With non-probability sampling, it is not possible to calculate the probability of a member of the population being selected for the sample and it is therefore difficult to assess the sample's reliability. However, certain population members may have no chance of being selected and there may be a danger of bias with particular types of individual being over-represented in the sample. Quota sampling is a form of non-probability sampling.

1.2.4 Sampling methods

(a) Simple random sampling (SRS)

In simple random sampling every member of the population has an equal probability of being selected for the sample. The selection procedure may simply involve drawing names from a hat. Alternatively, each member of the population appearing in the frame can be allocated a number and a computer can then be used to make a random selection from these numbers. It is usual to *sample without replacement*. This means that an individual who has been selected is not returned to the draw and therefore cannot appear more than once in the sample.

Simple random sampling is important in statistical theory and is a yardstick against which the efficiency of other sampling methods can be evaluated. Its main disadvantage is that it is often an impractical or expensive method to use when the population is large or geographically dispersed. It also requires a sampling frame.

(b) Stratified sampling

Where the population is made up of several distinct groups, for example, different social classes or age groups, simple random sampling does not guarantee that each group will be adequately represented in the sample. The sampling method can be improved by initially dividing the population into these groups (or strata) and then randomly sampling individuals from each stratum, a procedure known as stratified sampling.

Example

Age of employees at ABC Ltd	% of employees
under 21	10
21 to 40	50
41 and over	40

To take a stratified sample of fifty people from the above population one could randomly select five (i.e. 10 per cent of 50) from the under 21 age group, twenty-five from the 21 to 40 group and twenty from the 41 and over group. This ensures that the make-up of the sample, at least in terms of age, reflects that of the population.

Though stratified sampling can be a very reliable sampling method, it can be expensive or impractical when the population is large. It also· requires some prior knowledge of the population in order to be able to decide on the strata.

(c) Quota sampling

Like stratified sampling, quota sampling is an attempt to select a sample which reflects the different groups which make up the population. It involves giving interviewers quotas of people to interview. For example, in the ABC company above, an interviewer might be asked to interview five people from the under 21 group, twenty-five from the 21 to 40 group and twenty from the 41 and over group, as before. However, while in stratified sampling the individuals are sampled randomly from the strata, in quota sampling it is the *interviewer* who decides who should be included in the sample.

Quota sampling is thus a form of non-probability sampling since the chances of selection depend on the whim of the interviewer and who happens to be around when the survey is carried out. The sample may therefore be biased towards particular types of individual. However, quota sampling is relatively cheap and does not require a sampling frame.

(d) Systematic sampling

In systematic sampling every kth member of the sampling frame is selected after a random starting point. For example, if the sample is to be 1/20 of the population a random number (between 1 and 20) could be used to select the first name on a Register of Electors and thereafter every twentieth name could be chosen from the list. The chief advantage of the method is its simplicity. However, there is some danger of bias occurring if the sample reflects hidden periodicities in the sampling frame (e.g. if every twentieth name on a list of employees was a departmental head).

(e) Cluster sampling

Sometimes it may not be possible to obtain a sampling frame of all the individuals in the population. For example, it may be difficult to obtain a list of people employed in banking in a large city. However, a list of bank branches could be more easily obtained and a random sample of these could then be selected. At the selected branches all the employees would be interviewed. Since whole clusters of individuals (i.e. all the employees of a particular branch) are being selected, rather than the individuals themselves, this technique is known as cluster sampling.

Cluster sampling is generally relatively cheap since a large number of interviews can be carried out at one location. However, it is less reliable than simple random sampling. For example, employees at a particular bank branch may hold very similar opinions and, by sampling only a few branches, the full diversity of opinion in the population will not be represented in the sample.

(f) Multi-stage sampling

If the number of individuals in a selected cluster is too large for them all to be interviewed, multi-stage sampling can be used. This involves breaking the selected clusters down into smaller clusters and then taking a random sample of these smaller clusters. For example, to select a sample of students at polytechnics in England and Wales the following procedure could be used:

Stage 1   Select a random sample of polytechnics
Stage 2   For each polytechnic selected, take a random sample of courses
Stage 3   For each course selected, take a random sample of students

Like cluster sampling, multi-stage sampling has relatively low costs because interviews can be confined to a limited number of geographical areas. Also, a sampling frame of all the individuals in the population is not required. In the example above, initially, only a list of polytechnics would be required. It is, however, less reliable than simple random sampling.

(g) Sampling with probability proportional to size

In the sampling method described above, if polytechnic A has 10,000 students while polytechnic B has only 5000 then a student at A would have less chance of appearing in the final sample than a student at B. To avoid this, sampling with probability proportional to size can be used. This would mean that the chances of polytechnic A being selected in stage 1 are double those of polytechnic B. Similarly, at stage 2, the probability of selecting courses within the polytechnics could be weighted according to the number of students on each course. This would ensure that all polytechnic students had the same probability of being selected for the final sample.

1.2.5 Mail questionnaires and personal interviews

The actual collection of the data in a survey may involve sending out a mail questionnaire or conducting personal interviews. The main advantages of a mail questionnaire over personal interviews are

1   Mail questionnaires generally involve lower costs especially when a population is geographically dispersed
2   Replies to questions are not influenced by interviewers (i.e. there is no interviewer bias)
3   The respondent has time to give a considered answer to questions.

The main disadvantages are

1   Only a small, untypical group of people may bother to respond to the questionnaire
2   It may not be possible to establish exactly who completed the questionnaire (e.g. a manager may pass it on to his or her assistant)
3   Questions need to be straightforward
4   All the questions can be read before the respondent completes the

questionnaire. Therefore questions like 'Can you name a brand of coffee?' cannot be included if a later question asks 'Do you ever buy Nescafé coffee?'. Also, questions requiring spontaneous answers cannot be included.

1.2.6 Questionnaire design

A questionnaire should only include questions relevant to the objectives of the survey. The questions themselves should be short, clear and precise. Unambiguous and leading questions, which attract the respondent towards a particular answer should be avoided. Where possible, questions should include a complete list of possible responses and require the respondent to indicate his or her answer by ticking a box. This enables the results to be easily coded for computer analysis.

1.3 Presenting data

This topic does not appear on the CACA(2.6) syllabus

When presenting the results of a study or survey, long, verbal descriptions of statistical data can be extremely tedious and difficult to follow. This can be avoided by making careful use of tables, graphs and charts which will quickly convey the key features of the data. All tables, graphs and charts should have a clear title, an indication of the units being used (e.g. £s, kilograms, thousands of customers), labels for axes (or headings for rows and columns in a table) and, if secondary data is being used, a reference to its source.

1.3.1 Frequency tables (or distributions)

A table like Table 1.1 is known as a frequency table or frequency distribution.

Table 1.1

Length of service of employees at company XYZ (years)	Number of employees
0 to under 1	4
1 to under 2	10
2 to under 3	19
3 to under 4	8
4 to under 7	9
	Total 50

The categories into which the possible lengths of service have been divided (0 to under 1 year, 1 to under 2 years etc.) are known as *classes* and the number of employees belonging to a class is known as the *frequency* of the class (e.g. the frequency of the first class is 4). If the frequencies were

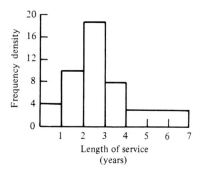

Figure 1.1 Length of service of employees at XYZ Ltd
Source: Personnel department

Many textbooks have no label for the vertical axis of a histogram while others label it 'frequency'

expressed as percentages of the total frequency, they would be known as *relative frequencies*. Thus the relative frequency of the first class is 8 per cent (i.e. $(4/50) \times 100$).

The width of each class is called the *class interval*. All the classes have a class interval of 1 year except the last class which has a class interval of 3 years.

1.3.2 Histograms

A histogram can be used to depict a frequency distribution. Figure 1.1 shows a histogram for the length of service data. Note that, for a histogram, it is the *area*, rather than the height, of the bars which is proportional to the frequency of the classes. This is important when a frequency distribution has unequal class intervals, as in this case. Because the last class is three times wider than the other classes, the height of its bar has been reduced to one third of its frequency. It therefore has a height of 3 (i.e. 9/3). Because the height of the bars does not measure frequency, the vertical axis of a histogram is labelled 'frequency density'.

1.3.3 Cumulative frequency distributions and ogives

A cumulative frequency distribution for the length of service data is shown in Table 1.2.

Table 1.2

Length of service of employees of company XYZ	Number of employees = cumulative frequency
under 1 year	4
under 2 years	14 (i.e. 4 + 10)
under 3 years	33 (i.e. 14 + 19)
under 4 years	41 (i.e. 33 + 8)
under 7 years	50 (i.e. 41 + 9)

This shows the number of employees with a length of service less than each of the upper class boundaries of the original frequency distribution.

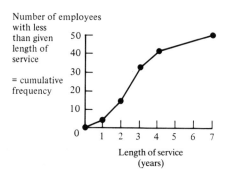

Figure 1.2 Length of service of employees at XYZ Ltd
Source: Personnel department

For example, a total of fourteen employees had less than two years' service: the four employees with less than one year plus the ten with between one and two years' service.

A graph of a cumulative frequency distribution is called an ogive and Figure 1.2 shows an ogive for the length of service data.

1.3.4 Pie charts

Pie charts (Figure 1.3) can be used to show the relative size of components which make up a total figure. Note that the angles at the centre of the chart are proportional to the size of the sectors. For example, if a sector represents 10 per cent of the total its angle at the centre will be 10 per cent

The horizontal axis of the ogive is the same as that for the histogram. Note that the points of the ogive have been plotted above the upper class boundaries not above the mid-points of the classes. Obviously, no employee will have less than zero years' service so the ogive initially has a height of zero.

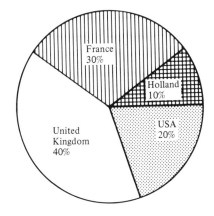

Figure 1.3 Analysis of XYZ's 1986 sales by country
Source: XYZ's sales department

of 360 (i.e. 36°). The effectiveness of pie charts is lost if they contain too many sectors. They are also less effective than bar charts when it is necessary to compare how different totals are made up.

The angles of the sectors should not be written on the pie chart.

1.3.5 Bar charts

Figure 1.4 shows two types of bar chart. The component bar chart enables totals to be compared and also shows how these totals are made up. If the total figures are of no interest, a multiple bar chart enables the size of the components to be compared.

1.3.6 Z charts

A Z chart can be used to monitor sales (or production) over a twelve-month period. It consists of three graphs: a graph of the actual sales for each month, a graph of the total sales for the year so far (cumulative sales) and a moving annual total (MAT) graph which gives the total sales for the twelve months up to and including the current month. The MAT graph is useful in

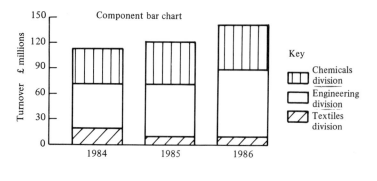

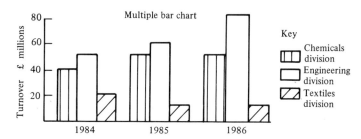

Figure 1.4   XYZ: Turnover of divisions 1984–6

determining the underlying trend in sales because it ignores short-term seasonal patterns.

Example

The monthly sales (number of units) of company X for 1985 and 1986 are given below. Draw a *Z* chart for 1986.

Month	Jan	Feb	Mar	Apr	May	Jun	Jul	Aug	Sep	Oct	Nov	Dec	Total
1985 Sales	8	10	14	15	18	20	30	46	18	11	10	7	207
1986 Sales	10	11	14	16	20	25	40	50	20	12	11	9	

For 1986 the cumulative sales and the MAT need to be calculated.

Month	Jan	Feb	Mar	Apr	May	Jun	Jul	Aug	Sep	Oct	Nov	Dec
Cumulative sales	10	21	35	51	71	96	136	186	206	218	229	238
MAT	209	210	210	211	213	218	228	232	234	235	236	238

For a given month the MAT has been calculated as follows:

$$\text{MAT} = \text{Total sales for preceding 12 months (or last MAT figure)}$$
$$+ (\text{Sales for the month} - \text{Sales in corresponding month}$$
$$\text{last year})$$

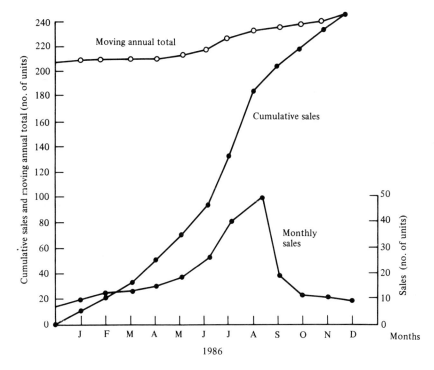

Figure 1.5 Z-chart for sales of Company X
Source: Sales office

For example

$$\text{MAT (for Jan 1986)} = 207 + (10 - 8) = 209$$
$$\text{MAT (for Feb 1986)} = 209 + (11 - 10) = 210 \text{ etc.}$$

The Z chart for this data is plotted in Figure 1.5. Because the actual sales figures are so much smaller than the MAT and cumulative figures they are plotted against a different axis. The chart shows that while the actual sales during 1986 first increase and then decline, probably as a result of seasonal factors, the underlying trend in sales (as indicated by the MAT graph) is upwards.

The previous year's figures do not appear on the Z chart and they are simply needed to calculate the moving annual totals.

The MAT for the last month on the Z chart always equals the last cumulative figure. This provides a useful check on your calculations.

1.3.7 Lorenz curves

A Lorenz curve is designed to show the degree of inequality in the way a total figure (e.g. income or wealth) is distributed among the population. For example, if total sales in an industry were equally distributed among the firms in that industry then 10 per cent of the firms would have 10 per cent of the sales, 20 per cent of the firms would have 20 per cent of the sales and so on. However, if only 10 per cent of the firms accounted for 90 per cent of the sales this would suggest a great inequality in the distribution of sales.

Example

Draw a Lorenz curve for the data in Table 1.3 which relates to the wages paid to the employees of a company.

Table 1.3

Gross wage per week	Number of employees	Total wages paid to each group of employees
under £60	50	£2,500
£60 to under £100	80	£6,400
£100 to under £120	50	£5,000
£120 and over	20	£3,100
	Total 200	Total wage bill £17,000

To draw a Lorenz curve the percentage of employees in each wage bracket and the percentage of the total wage bill being paid to each group of employees must first be calculated. These percentages are then cumulated and it is the cumulative percentages which are plotted on the graph (Figure 1.6).

Table 1.4

No. of employees	% of total employees	Cumulative %	Total wages	% of total wage bill	Cumulative %
50	25%	25%	£2,500	14.7%	14.7%
80	40%	65%	£6,400	37.7%	52.4%
50	25%	90%	£5,000	29.4%	81.8%
20	10%	100%	£3,100	18.2%	100.0%
200	100%		£17,000	100.0%	

The calculations above show that the lowest paid 25 per cent of employees earn only 14.7% of the total wages paid out each week by the company. If the total wages paid out were evenly distributed amongst the workforce then the Lorenz curve would have followed the line of equal distribution on the graph. The greater the inequality in the distribution, then the greater will be the deviation of the Lorenz curve from the line of equal distribution.

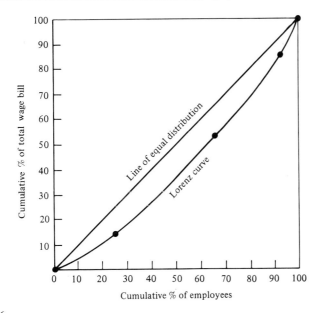

Figure 1.6

2
Measures of location and dispersion

The material in this chapter is relevant to the following syllabuses: CIMA, ICAEW, ICAS, ICAI, CACA(1.5), ICSA

2.1 Introduction

In the previous chapter it was shown that a frequency distribution or histogram can be used to represent the pattern occurring in a set of data. However, rather than listing a complete frequency distribution, it is sometimes preferable simply to state a single value which is representative or typical of the original figures. Such a value is known as an average or a measure of location.

Measures of location are also called measures of central tendency.

A second group of measures, known as measures of dispersion, can be used to describe the amount of variation in a set of data. For example, they could be used to describe the extent to which output at a factory varies from day to day or how much variation there is in the size of orders received by a mail-order company.

2.2 Sigma notation

When the Greek letter Σ (sigma) appears in a formula it means 'add up'. Thus Σx means 'add up all the figures in the column or row labelled x'. Note that, in the absence of brackets, adding up is always the last operation performed. Thus Σx^2 means 'square all the xs first and then add up the resulting figures' while Σxy would mean 'multiply the corresponding xs and ys together and then add up the results'.

Example

Given x 1 2 0 3
         y 4 1 6 2

Find (a) Σx (b) Σx^2 (c) Σxy (d) $(\Sigma x)^2$

(a) $\Sigma x = 1 + 2 + 0 + 3 = 6$
(b) $\Sigma x^2 = 1^2 + 2^2 + 0^2 + 3^2 = 1 + 4 + 0 + 9 = 14$
(c) $\Sigma xy = (1 \times 4) + (2 \times 1) + (0 \times 6) + (3 \times 2) = 12$
(d) $(\Sigma x)^2 = (6)^2 = 36$

2.3 Measures of location

The main measures of location are the arithmetic mean, the median and the mode. All of these measures should be used carefully since it is possible to derive results which are neither typical nor representative of the data from which they have been calculated. Sometimes it may be necessary to use more than one measure in order adequately to describe the original data.

2.3.1 The arithmetic mean

The arithmetic mean (or simply the mean) is the 'average' familiar to most people. It involves dividing the total of a set of figures by the number of figures. For example the mean of 2, 6 and 7 is

$$\frac{2+6+7}{3} = 5$$

Normally the Greek letter μ (mu) is used to represent the mean of a population while the mean of a sample is denoted by $\bar{x}$ (x bar). Thus the process of calculating the mean can be represented by the following formula

$$\bar{x} = \frac{\Sigma x}{n} \quad \text{(where } x \text{ represents the figures to be averaged and } n = \text{ the number of figures)}$$

2.3.2 Finding the mean from a frequency distribution (grouped data)

A frequency distribution for the weekly sales of a product is given below.

Sales per week (no. of units)	No. of weeks
0 to under 10	4
10 to under 20	20
20 to under 40	16
40 to under 60	8
60 to under 100	2
	Total 50

Because the exact sales are not given (e.g. two of the weeks had sales of somewhere between sixty and 100 units, but the exact sales are unknown) it is necessary to assume that the weekly sales are equivalent to the mid-points of the classes. The formula for finding the mean from a frequency distribution is therefore

$$\bar{x} \text{ or } \mu = \frac{\Sigma fx}{\Sigma f} \quad \text{where } x = \text{ the mid-point of each class and } f = \text{ the frequency of each class}$$

For the above distribution the calculations are set out as follows

Sales per week	f	x = class mid-points	fx
0 to under 10	4	5	20 (i.e. 4 × 5)
10 to under 20	20	15	300 (i.e. 20 × 15) etc.
20 to under 40	16	30	480
40 to under 60	8	50	400
60 to under 100	2	80	160
	$\Sigma f = 50$		$\Sigma fx = 1360$

This result could be interpreted as mean sales of 272 units every ten weeks.

Therefore the mean sales $= \dfrac{1,360}{50} = 27.2$ units per week.

2.3.3 Features of the mean

The mean is the most widely used measure of location and it has mathematical properties which make it useful in advanced statistical work. Sometimes the mean is an 'impossible' value (e.g. in the sales example it may not be possible to sell 0.2 units). However, its most serious weakness is that it can be distorted by extreme values. For example, suppose that the weekly incomes of a random sample of five people were £90, £100, £110, £120, £2000. The mean income is £484 which is not typical or representative of any of the incomes.

2.3.4 The median

The median is the middle value in a set of figures when the figures have been placed in ascending order.

Example
The value of five orders received by a company is shown below. Find the median value of the orders.

£800, £600, £100, £760, £550

Arranging the figures in ascending order gives

£100, £550, £600, £760, £800

Therefore the median is £600.

When there is an even number of figures the median is halfway between the middle two figures. For example, the median of £60, £90, £100, and £150 is £95.

2.3.5 Finding the median from a frequency distribution (grouped data)

An ogive can be used to find the median from a frequency distribution. Figure 2.1 shows the ogive for the sales data (given in Section 2.3.2). The median is the level of sales which has a cumulative frequency equal to half

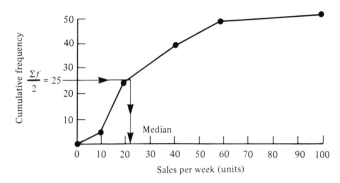

Figure 2.1

of the total frequency (i.e. $50/2 = 25$). From the graph it can be seen that the median level of weekly sales is approximately twenty-one units.

2.3.6 Features of the median

Unlike the mean, the median is not distorted by extreme values. Thus, for the weekly incomes of the five people mentioned in Section 2.3.3 (£90, £100, £110, £120, £2000) the median is £110. It can also be found from a frequency distribution with open-ended classes without having to make arbitrary assumptions about the width of these classes. However, the median does not fully take into account every value in a set of data and it is less useful than the mean in advanced statistical work.

2.3.7 The mode

The mode is the most frequently occurring value in a set of data. For example, if the following figures represent estimates of the rate of inflation for the coming year by nine economists, the mode is 5 per cent.

2% 5% 1% 5% 3% 5% 5% 1% 5%

Some sets of data have two modes and are said to be *bimodal* (e.g. 2, 6, 6, 6, 7, 8, 9, 9, 9, 10 has modes of 6 and 9).

2.3.8 Estimating the mode from a frequency distribution (grouped data)

If data has been organized into a frequency distribution an estimate of the mode can be found by drawing a histogram. Figure 2.2 shows a histogram of the sales data. The class with the highest bar on the histogram (in this case the 10 to under 20 units class) is called the *modal class*. To estimate where the mode lies within this class, two lines are drawn between the top corners of the highest bar and the top of adjacent bars, as shown. The mode lies directly below the point of intersection of these lines. For the sales data, the modal level of sales is about sixteen units. However, this is only a rough estimate and often it may be sufficient simply to state the modal class.

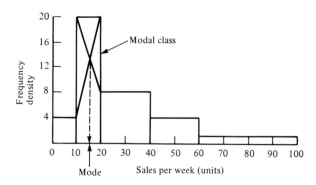

Figure 2.2

2.3.9 Features of the mode

The mode is very simple to calculate and, unlike the mean, it cannot result in 'impossible' values and it is not distorted by extreme values. However, for it to have any meaning one value (or for bimodal distributions two values) must occur much more frequently than others.

2.3.10 Weighted averages

Sometimes numbers which are to be averaged have different degrees of importance. For example, a student's examination mark may be considered to be more important than his or her coursework mark or, when measuring the rate of inflation, a rise in the price of bread may be considered to be more important than a rise in the price of fur coats. In these cases weights are attached to each number as a measure of their importance and a weighted average is calculated using the following formula

$$\text{weighted average} = \frac{\Sigma xw}{\Sigma w} \quad \begin{array}{l}\text{where } x = \text{the original values} \\ \text{and } w = \text{the weights}\end{array}$$

Example

A job applicant takes tests in English, Mathematics and Computer Programming. The applicant's scores on the tests, and the weightings attached to each test, are shown below. Calculate the weighted average score.

Test	$x = test\ score$	$w = weight$	xw
English	20	3	60 (i.e. 20×3)
Maths	50	5	250
Computing	70	2	140
		$\Sigma w = 10$	$\Sigma xw = 450$

Therefore the weighted average score $= \dfrac{450}{10} = 45$

2.4 Measures of dispersion

It may not be enough to describe a set of data using only a measure of location. For example, a machine producing a product may be achieving the desired mean weight of 100 grams. However, the individual weights of the items produced may vary from 50 to 150 grams so that only a small part of the output is satisfactory. To detect this it is necessary to measure the variability or dispersion of the data. The main measures of dispersion are the range, the interquartile range, the mean deviation and the standard deviation.

2.4.1 The range

The range is the simplest measure of dispersion and is defined as the highest number in a set of data – lowest number. Thus the range of the numbers 1, 6, 7, 8, 9, 10 is 9 (i.e. $10 - 1$).

The range is of limited use because it ignores all except the highest and lowest values and it can therefore also be distorted by extreme values. For example, the numbers 1, 3, 5, 7, 9, 11 and 1, 5, 5, 5, 5, 11 both have a range of 10 but they have very different patterns of variation. However, because of the speed with which it can be calculated, the range is used in quality control.

2.4.2 The interquartile range

The interquartile range (IQR) avoids the problem of extreme values by ignoring the smallest 25 per cent and largest 25 per cent of the numbers in a set of data. To calculate the IQR it is first necessary to identify the quartiles.

The *lower quartile* (Q_1) is the value below which 25 per cent of observations fall. It therefore has a cumulative frequency equal to one quarter of the total frequency. For the sales data, Q_1 will have a cumulative frequency of 50/4 i.e. 12.5. Figure 2.3 shows that Q_1 equals about 14 units.

The *upper quartile* (Q_3) is the value below which 75% of the observations fall and it therefore has a cumulative frequency equal to 3/4 of the total

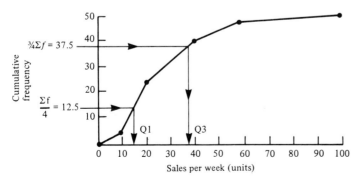

Figure 2.3

frequency (i.e. 37.5 for the sales data). Figure 2.3 suggests that Q_3 is about 37 units.

The interquartile range is defined as $Q_3 - Q_1$. Thus for the sales data the IQR is $37 - 14 = 23$ units. The IQR therefore measures the spread of the 'middle 50 per cent' of weekly sales. However, like the range, the IQR does not fully take into account all the values in set of data.

The *semi-interquartile range* (or quartile deviation) is simply half of the IQR (i.e. 11.5 units for the sales data). It shows the average distance of the quartiles from the median.

2.4.3 The mean deviation

The mean deviation is calculated as follows
1   Subtract the mean from every number (i.e. calculate $x - \bar{x}$). The resulting figure is called a deviation.
2   Remove any minus signs from the deviations (this gives $|x - \bar{x}|$ where $||$ means ignore any minus sign).
3   Sum these 'modified' deviations (i.e. calculate $\Sigma |x - \bar{x}|$).
4   Divide by the number of observations (n).

Thus the mean deviation $= \dfrac{\Sigma |x - \bar{x}|}{n}$

Example
The following figures represent the production achieved in a factory over a period of four days: 20 units, 25 units, 15 units, 12 units. Calculate the mean deviation.

Let $x =$ the production levels. Therefore $\bar{x} = 18$ units.

x	$x - \bar{x}$	$|x - \bar{x}|$
20	2	2
25	7	7
15	−3	3
12	−6	6
	0	$\Sigma|x - \bar{x}| = 18$

Note that the sum of the deviations ($\Sigma (x - \bar{x})$) always equals zero and this is the reason for ignoring the minus signs in the mean deviation and squaring the deviations in the standard deviation. Nevertheless, as a check on calculations, it is always worth adding up the deviations column to make sure that it does sum to zero.

Therefore the mean deviation $= 18/4 = 4.5$ units.

The mean deviation is simple to calculate and interpret (it represents the average 'distance' of each number from the mean). However, because the minus signs are ignored, it is a difficult measure to analyse mathematically and is therefore less widely used than the standard deviation.

2.4.4 The standard deviation

There are two symbols for sigma in Greek. Σ is capital sigma.

The standard deviation is the most important measure of dispersion. Normally the Greek letter σ (sigma) is used to represent the standard deviation of a population while the letter s denotes the standard deviation of a sample. Calculating the standard deviation involves the following steps

1 Calculate the deviations (as for the mean deviation) (i.e. calculate $x - \bar{x}$).
2 Square all the deviations i.e. calculate $(x - \bar{x})^2$.
3 Sum the squared deviations i.e. calculate $\Sigma(x - \bar{x})^2$.
4 Divide this total by the number of figures in the set of data less 1 i.e.

    calculate $\dfrac{\Sigma(x - \bar{x})^2}{n - 1}$

5 Find the square root of the above figure.

Thus the standard deviation $= \sqrt{\dfrac{\Sigma(x - x)^2}{n - 1}}$

You may previously have met the formula

$$\sqrt{\dfrac{\Sigma(x - \bar{x})^2}{n}}$$

Example

Given the following weights (in ounces) of jars of coffee which have been filled by a machine, find the standard deviation.

    18, 19, 19, 21, 18

The mean weight, $\bar{x} = 19$ oz and n, the number of figures $= 5$. Thus:

The formula used here (which is the one given to CIMA and CACA students in the examination) is really designed to be used with data from a sample. It has been 'corrected' to take into account the fact that the standard deviation of a sample tends to underestimate the degree of variation in the population.

$x = weights$	$x - \bar{x} = deviations$	$(x - \bar{x})^2 = squared\ deviations$
18	-1	1
19	0	0
19	0	0
21	2	4
18	-1	1
$\Sigma(x - \bar{x}) = 0$	$\Sigma(x - \bar{x})^2 = 6$	

Therefore the standard deviation $= \sqrt{\dfrac{6}{5 - 1}} = \sqrt{1.5} = 1.22$ oz

The square of the standard deviation is known as the *variance*. This is normally represented by either σ^2 or s^2. The variance of the above data is clearly 1.5 oz^2.

'Ounces squared' may seem a strange unit, but it is technically correct because in the calculation of the variance all units are squared.

2.4.5 Calculating the standard deviation from a frequency distribution (grouped data)

The formula for calculating the standard deviation from a frequency distribution is

standard deviation $= \sqrt{\dfrac{\Sigma fx^2}{\Sigma f} - \left(\dfrac{\Sigma fx}{\Sigma f}\right)^2}$

where f = frequency of the classes and x = class mid-points.

For the sales data the calculations are set out as follows

Note that Σfx^2 does not equal $(\Sigma fx)^2$. In the first expression only the x is squared. Also note that $fx^2 = fx(x)$. Therefore to calculate fx^2 the fx column is multiplied by the x column.

Sales per week	f	x = mid-points	fx	fx²
0 to under 10	4	5	20	100
10 to under 20	20	15	300	4,500
20 to under 40	16	30	480	14,400
40 to under 60	8	50	400	20,000
60 to under 100	2	80	160	12,800
	$\Sigma f = 50$		$\Sigma fx = 1,360$	$\Sigma fx^2 = 51,800$

The calculations for Σf and Σfx have, of course, already been carried out in the calculation of the mean.

Therefore the standard deviation $= \sqrt{\dfrac{51,800}{50} - \left(\dfrac{1,360}{50}\right)^2}$

$= \sqrt{1,036 - 739.84}$

$= 17.2 \text{ units}$

2.4.6 Features of the standard deviation

The standard deviation, unlike the range and interquartile range, takes every value in a set of data fully into account. Because of its mathematical properties it is important in later statistical work. However, it is a measure which is difficult to interpret (though clearly the larger it is, then the more variation there is in the data) and it tends to be over-influenced by extreme values because it involves squaring the deviations.

2.4.7 The coefficient of variation

To compare the variation of two or more sets of data, it is often useful to calculate the relative variation. For example, suppose that an investment analyst is comparing two investments A and B which both generate annual returns which vary such that they both have a standard deviation of £20. This variation would be much more noticeable for investment A, which has a mean annual return of only £60, than for B, which has a mean annual return of £500. The coefficient of variation would reflect this difference.

$$\text{Coefficient of variation} = \frac{\text{standard deviation}}{\text{mean}} \times 100$$

Therefore for A the coefficient of variation $= 20/60 \times 100 = 33.3\%$; and for B the coefficient of variation $= 20/500 \times 100 = 4.0\%$.

3
Time series and index numbers

The material in this chapter is relevant to the following syllabuses, except where otherwise indicated: CIMA, ICAEW, ICAS, ICAI, CACA(1.5), CACA(2.6), CIPFA, ICSA

3.1 Introduction

Data recorded at intervals over a period of time (e.g. records of monthly sales figures or the number of employees absent per day) is called a time series. Time series analysis refers to the process of decomposing the data into a number of components in order to establish if a pattern exists. If a pattern can be discerned then this may be particularly useful when forecasts need to be made.

Index numbers are useful when large sets of data (e.g. the prices paid for goods and services in an economy) need to be compared, normally between different points in time.

3.2 Time series analysis

This topic does not appear on the CACA(1.5) syllabus

3.2.1 The components of a time series

In most time series some, or all, of the following components can be observed

1   A trend (T). This represents the long-term underlying movement in the time series.
2   A seasonal pattern (S).
3   A long-term cycle (C) which often reflects the swings in the economy from boom years to years of depression. Since this component is often very long-term in nature, its measurement will not be considered here.
4   Small irregular movements reflecting the many unpredictable factors which affect a time series. These constitute the residual component (R).

3.2.2 Models of time series

The additive time series model assumes that any observation in the series (Y) is equal to the sum of the four components i.e. $Y = T + S + C + R$.

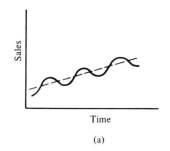

Sales

Time

(a)

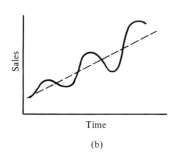

Sales

Time

(b)

Figure 3.1

The multiplicative model assumes that an observation is equal to the product of the components i.e. $Y = T \times S \times C \times R$. Generally the additive model should be used when the seasonal swings in the graph remain fairly constant as the trend increases (Figure 3.1(a)) while the multiplicative model should be used where the swings become more marked as the trend increases (Figure 3.1(b)).

3.2.3 Analysing a time series using the additive model

Consider the sales figures given in column 1 of Table 3.1. These have been plotted on the graph in Figure 3.2. The stages in analysing this time series using the additive model are as follows

Table 3.1

Year	Qtr	(Col. 1) Sales (units)	(Col. 2) Moving average	(Col. 3) Centred average	(Col. 4 = Col. 1 – Col. 3) Seasonal deviation = sales – centred average
1984	1	28			
	2	44			
	3	50	33	33.5	+16.5
	4	10	34	35	−25
1985	1	32	36	38	−6
	2	52	40	41	+11
	3	66	42	43.5	+22.5
	4	18	45	46	−28
1986	1	44	47	47.5	−3.5
	2	60	48	49.5	+10.5
	3	70	51		
	4	30			

	Quarter 1	Quarter 2	Quarter 3	Quarter 4	
1984			+16.5	−25.0	
1985	−6.0	+11.0	+22.5	−28.0	
1986	−3.5	+10.5			
					Total
Mean	−4.75	+10.75	+19.5	−26.5	−1
Adjustment	+0.25	+0.25	+0.25	+0.25	
Mean Seasonal Deviations	−4.5	+11.0	+19.75	−26.25	

If the data was monthly, a twelve-month moving average would normally be used.

Note the moving averages are plotted on the graph halfway between the quarters i.e. in the same positions they appear in on the table. The centred averages fall on the trend line halfway between the moving averages i.e. directly above the quarters. A clear disadvantage of the use of moving averages to measure the trend is that they leave a gap at the start and end of the graph.

1   Calculate the trend using a four-quarterly moving average (column 2 in the table). The first moving average is the mean of the sales for the four quarters of 1984 (i.e. 28, 44, 50, 10). Since this relates to the middle of the four periods it is written halfway between the middle two periods. To calculate the next moving average the sales figure for 1984 quarter 1 is 'deleted' and the figure for 1985 quarter 1 is added to the numbers to be averaged so that the mean of 44, 50, 10 and 32 is found. This process of moving down a quarter at a time is repeated until the last four sales figures have been averaged. The moving averages are plotted on the graph and show that the trend in sales is upwards.

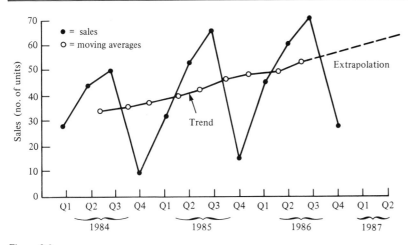

Figure 3.2

2   Because the moving averages have been written halfway between the sales figures it is not possible to compare the actual sales with the trend. Therefore the centred averages (column 3) are calculated next. These are the values midway between the pairs of moving averages. Thus the first centred average is $(33+34)/2 = 33.5$, the second is $(34+36)/2 = 35$ and so on.

3   Calculate the seasonal deviations (column 4) by subtracting the centred averages from the actual sales figures. Thus the first deviation is $50 - 33.5 = +16.5$ showing that sales for quarter 3 of 1984 were 16.5 units above the trend.

4   Put the deviations from column 4 into a new table which will enable the mean seasonal deviation for each quarter to be calculated.

5   The mean seasonal deviations should sum to zero. In this case the sum is -1 so 0.25 is added to each figure. The resulting figures show that quarter 3 is the peak season, with sales on average being 19.75 units above the trend, and quarter 4 is the slack period, with sales on average 26.25 units below the trend.

3.2.4 Making a forecast using the additive model

If a sales forecast is required for 1987 quarter 1 this can be obtained in two stages

1   The trend line on the graph is extrapolated into the future. There are several ways of doing this. One method involves fitting a regression line to the data (see Chapter 9). The simplest method is to use your judgement to extend the trend line, and read off the appropriate value. Looking at Figure 3.2, the trend forecast for 1987 quarter 1 appears to be about 57 units.

2   Adjust the forecast to take into account the season. The average seasonal deviation for quarter 1 suggests that its sales would be

expected to be 4.5 units below the trend. Thus the sales forecast is: $57 - 4.5 = 52.5$ units.

Of course, this forecast assumes that the steadily increasing trend and the observed seasonal pattern will continue into the future. In particular, no account has been taken of possible external factors such as a competitor launching a major sales promotion campaign or the government increasing the tax on the product.

3.2.5 Deseasonalizing data using the additive model

In order to compare figures on an equal basis it is often necessary to deseasonalize them. For example, in the UK unemployment may decrease in the summer. However, in order to detect whether the fall represents a real improvement the figures have to be seasonally adjusted. If the additive model has been used, figures can be adjusted as follows:

$$\text{Deseasonalized figure} = \text{Original value} \\ - \text{average seasonal deviation}$$

For example, deseasonalizing the sales figure for 1984 quarter 1:

$$\text{Deseasonalized sales} = 28 - (-4.5) = 32.5 \text{ units}$$

while for 1984 quarter 2, deseasonalized sales $= 44 - (11) = 33$ units.

3.2.6 Using the multiplicative time series model

Analysing a series using the multiplicative model involves the following steps

1 The moving and centred averages are calculated exactly as before.
2 This time the seasonal variation is measured by seasonal indices where, for a given period

$$\text{Seasonal index} = \frac{\text{actual sales}}{\text{centred average}} \times 100$$

Thus for 1983 quarter 4 the seasonal index $= (50/33.5) \times 100 = 149.3$ which shows that sales in this period were 49.3 per cent higher than the trend. These figures would replace those in column 4 in Table 3.1 and would then be averaged like the seasonal deviations in a new table. This time the sum of the average seasonal indices should be 400. If this sum is not obtained each index can be multiplied by (400/actual total). For the data in Table 3.1 the average seasonal indices are: quarter 1: 89.1 per cent, quarter 2: 125 per cent, quarter 3: 151.7 per cent and quarter 4: 34.2 per cent. This shows for example that, on average, sales in quarter 4 are only 34.2 per cent of the trend.

To forecast sales using the multiplicative model the trend is forecast as before and then the sales forecast is obtained as follows

$$\text{Sales forecast} = \text{trend forecast} \times \frac{\text{appropriate average seasonal index}}{100}$$

If monthly data is being used the sum should be 1200.

Thus for 1987 quarter 1 the sales forecast $= 57 \times 89.1/100 = 50.8$ units.

To deseasonalize data using the multiplicative model the following formula is used

$$\text{Deseasonalized figure} = \frac{\text{original value}}{\text{appropriate average seasonal index}} \times 100$$

Thus, the deseasonalized 1984 quarter 1 sales $= (28/89.1) \times 100 = 31.4$ units.

3.2.7 Exponential smoothing

This topic does not appear on the CIMA, ICAEW, ICAI, or ICSA syllabuses

Exponential smoothing is a short-term forecasting method which, in its simplest form, is applicable when there is no trend or seasonality present in the data. Forecasts are made using the following formula

Forecast for next period $=$ last forecast $+ \alpha$ (error in last forecast)

where α (alpha), which is known as the smoothing constant, always has a value between 0 and 1. The forecast error is defined as

Actual value $-$ forecast

Table 3.2 below shows the calculations for exponential smoothing, using an α value of 0.5. The forecast of 40 for period 1 was a rough estimate to start the process off.

Table 3.2

Period	(Col. 1) Actual sales	(2) Forecast	(3) = (1) − (2) Error	(4) $\alpha \times$ error	(5) = (2) + (4) Forecast for next period
1	50	40	+10	+5	45
2	41	45	−4	−2	43
3	45	43	+2	+1	44
4		44			

It can be shown that a forecast based on exponential smoothing is in fact a weighted average of all the previous observations: the more recent an observation is, then the higher will be its weight. The accuracy of the forecasts can be assessed by calculating the mean square error (MSE) where

$$\text{MSE} = \frac{\Sigma(\text{error})^2}{\text{no. of errors}}$$

Thus for the data above

$$\text{MSE} = \frac{(10)^2 + (-4)^2 + (2)^2}{3} = 40$$

Obviously, in practice, three forecasts is far too few to give an idea of the accuracy of the forecasting method.

Generally, the value of α which gives the lowest MSE is the one used to make the forecasts. Exponential smoothing has a number of advantages over other methods. These include its simplicity and low computer storage requirements, which are important when a large number of forecasts need to be made.

3.3 Index numbers

This subject does not appear on the ICAS, CIPFA or CACA(2.6) syllabuses

Index numbers are useful for monitoring variables like prices, wages and costs, normally over periods of time. One period is designated as the base period and the index for this period is set equal to 100. Index numbers for all other periods are then calculated relative to the base period. For example an index number for production costs is shown below with 1983 as base year

Year	1983	1984	1985	1986
Production cost index	100	90	110	115

This shows that production costs in 1984 were only 90 per cent of those in 1983. However, 1985's costs were 10 per cent higher than 1983's and 1986's costs were 15 per cent higher than those of 1983.

Unfortunately, the percentage change between two index numbers where neither is the base year cannot be found by simply subtracting one index from another. For example, the percentage rise in costs between 1985 and 1986 is not $115 - 110 = 5\%$. The correct percentage rise is

$$\frac{115 - 110}{110} \times 100 = 4.5\%$$

(i.e. $\dfrac{the\ increase\ or\ decrease}{earlier\ index} \times 100$)

3.3.1 Changing the base of an index number

Sometimes it is necessary to change the base of an index number. For example, to compare wages and prices, where the index numbers for these variables have different base years, one of the index numbers would have to have its base year changed. To change the base, simply identify the index's value at the year which is to become the new base, divide this into each number in the series and multiply by 100.

Example
Change the base of the price index below to 1985.

Year	1983	1984	1985	1986
Price index	100	110	125	130

Index with 1985 as base

$$1983: \frac{100}{125} \times 100 = 80 \qquad 1984: \frac{110}{125} \times 100 = 88$$

$$1985: \frac{125}{125} \times 100 = 100 \qquad 1986: \frac{130}{125} \times 100 = 104$$

3.3.2 Using index numbers for deflation

Index numbers are often used to assess whether variables like costs or wages are increasing in 'real terms' (i.e. when the effect of inflation has been removed). Given below are the weekly wages of a group of workers for the years 1983–5 and the values of a price index for these years.

Year	1983	1984	1985
Weekly wage	£90	£100	£110
Price index	100	120	130

To assess whether the purchasing power of the wages has increased, they are deflated. This involves dividing each wage by the index for that year and multiplying by 100. Thus the deflated wages are:

$$1983: \frac{90}{100} \times 100 = £90 \qquad 1984: \frac{100}{120} \times 100 = £83.33$$

$$1985: \frac{110}{130} \times 100 = £84.62$$

Therefore, the real value of the wages has declined between 1983 and 1985 because they failed to keep up with inflation. Note that the deflated wages are expressed in terms of the value of money in 1983, the base year.

3.3.3 Relatives

If an index number measures the change in price of only one item it is known as a price relative and is calculated as follows

$$\text{Price relative} = \frac{p_n}{p_o} \times 100$$

where p_n is the price in the current year and p_o is the price in the base year.

For example, suppose that electricity cost 5 pence per unit in 1985 and 6 pence in 1986. If 1985 is the base year (this is often stated as 1985 = 100) then the price relative for 1986 is $(6/5) \times 100 = 120$ which shows a 20 per cent increase in the price of electricity.

3.3.4 Weighted average of price relatives index numbers

Most useful index numbers measure the average movement in the prices of more than one item. However, taking a simple average of the price relatives would ignore the fact that some items are more important than others (e.g. potatoes may be more important than caviare). To take this into account weights are assigned to the items which reflect their relative importance. (In the RPI the weights reflect the percentage of total expenditure allocated to each item by the 'average' household.) A weighted average of price relatives index number can then be calculated as follows

$$\text{Index number} = \frac{\Sigma \text{price relative} \times \text{weight}}{\Sigma \text{weights}}$$

Example

Calculate a weighted average of price relatives index number for 1986 from the data below using 1985 as base year.

Commodity	1985 price	1986 price	Weight
A	£20 per ton	£30 per ton	30
B	£2 per litre	£1.80 per litre	20
C	£5 per dozen	£6 per dozen	50
			100

The price relatives for 1986 are, for A: $(30/20) \times 100 = 150$; for B: $(1.80/2) \times 100 = 90$ and for C: $(6/5) \times 100 = 120$. Thus the 1986 index number (with $1985 = 100$) is given by

$$\frac{(150 \times 30) + (90 \times 20) + (120 \times 50)}{100} = \frac{12,300}{100} = 123$$

which suggests a 23 per cent increase in prices.

3.3.5 Laspeyre and Paasche index numbers

A *weighted average of price relatives (WAPR) index should never be quantity weighted. For example, a transport company will probably buy more paper clips than delivery vans. If the quantities of the items bought were used as weights in a WAPR index, this would mean that a rise in the price of paper clips would affect the index more than a rise in the price of vans.*

One way of assessing the relative importance of items is to use the quantity purchased as a weight. A Laspeyre index uses the quantities of items purchased in the base year as weights and is calculated using the following formula

$$\text{Laspeyre index} = \frac{\Sigma p_n q_o}{\Sigma p_o q_o} \times 100$$

where p_n = price in current year, p_o = price in base year and q_o = quantity of item purchased in base year.

Note that the denominator of the index is the total expenditure on items in the base year ($\Sigma p_o q_o$). The numerator represents the expenditure which would have been incurred if the base year quantities of items had been bought, but at current year prices ($\Sigma p_n q_o$).

A Paasche index uses the quantities of items purchased in the current year as weights and is calculated as follows

$$\text{Paasche index} = \frac{\Sigma p_n q_n}{\Sigma p_o q_n} \times 100$$

where q_n = quantity of items purchased in the current year.

This index therefore compares the total expenditure on items in the current year ($\Sigma p_n q_n$) with the expenditure which would have been incurred if the same quantities of goods had been purchased, but at base year prices ($\Sigma p_o q_n$).

Example

Using 1985 as base year calculate a Laspeyre and Paasche index for 1986 from the following data

Commodity	1985		1986	
	Price (p_o)	Quantity purchased (q_o)	Price (p_n)	Quantity purchased (q_n)
A	£20 per ton	100 tons	£30 per ton	90 tons
B	£2 per litre	500 litres	£1.80 per litre	600 litres
C	£5 per dozen	50 dozen	£6 per dozen	80 dozen

$1985 = 100$. Thus

$$\Sigma p_n q_o = (30 \times 100) + (1.80 \times 500) + (6 \times 50) = £4,200$$

$$\Sigma p_o q_o = (20 \times 100) + (2 \times 500) + (5 \times 50) = £3,250$$

Therefore the Laspeyre index for 1986 is

$$\frac{4,200}{3,250} \times 100 = 129.2$$

$$\Sigma p_n q_n = (30 \times 90) + (1.80 \times 600) + (6 \times 80) = £4,260$$

$$\Sigma p_o q_n = (20 \times 90) + (2 \times 600) + (5 \times 80) = £3,400$$

Therefore the Paasche index for 1986 is

$$\frac{4,260}{3,400} \times 100 = 125.3$$

Because the Paasche index uses quantities of items purchased in the current year, it is more up to date than the Laspeyre index. However, unlike the Laspeyre, it requires data on the quantities of items purchased to be obtained for each new year, which can be expensive if a large number of items are involved. Also, since the weighting system changes each year, a Paasche index, unlike a Laspeyre, can really only be compared with the base year and not with any intervening year.

3.3.6 Practical problems of constructing index numbers

The main practical problems encountered when constructing index numbers are

1 When there are millions of goods and services only a representative sample of them can be included in the index.

2   Over time the quality of products may change so that the price of a product in two different periods may not be directly comparable.

3   It is often difficult to choose the base period which, ideally, should be a normal and recent period. This problem can sometimes be resolved by using a chain base index which always uses the previous period as base (e.g. the 1986 index would have 1985 as base, the 1987 index would have 1986 as base and so on).

4
Probability theory and decision analysis

The material in this chapter is relevant to the following syllabuses: CIMA, ICAEW, ICAS, ICAI, CACA(1.5), CACA(2.6), CIPFA, ICSA

4.1 Introduction

Most decisions involve an element of uncertainty. In particular, when decisions are based on information obtained from samples there can be no certainty that the information is perfectly accurate. In these circumstances probability theory can often be used to clarify the risks associated with particular courses of action.

This chapter considers the basic ideas of probability and the application of these ideas in decision making. Note that $p(x)$ will be used as shorthand for 'the probability of event x occurring'.

4.2 Sample space and Venn diagrams

The sample space is the set of all the possible outcomes which can occur if a sample is to be selected from a population or an experiment is to be carried out. For example, if a coin is to be tossed once then the sample space will simply consist of the two outcomes: head and tail. If the coin is tossed twice there are four possible outcomes in the sample space (where H = head and T = tail): H followed by H, H followed by T, T followed by H, T followed by T.

Sample space can be represented on a Venn diagram (Figure 4.1) (these can be useful in clarifying awkward probability problems). The circled outcomes on the Venn diagram show that outcomes can be grouped into events, in this case the event 'at least one head'.

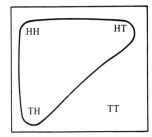

Figure 4.1

4.3 Three approaches to probability

(a) The a priori approach
The *a priori* definition of probability is

$$p(A) = \frac{\text{number of outcomes which represent the occurrence of event } A}{\text{total number of possible outcomes}}$$

The term 'a priori' is used because (unlike the empirical approach below) this approach allows a probability to be calculated before any experience has been gained of the event's frequency of occurrence.

where all of the outcomes are assumed to be equally likely. It follows from this definition that all probabilities must be between 0 (impossible) and 1 (certain).

Example

A coin is tossed twice. Find the probability of obtaining at least one head.

Three outcomes represent the occurrence of at least one head and there is a total of four possible outcomes (see Figure 4.1). Therefore p(at least 1 head) = 3/4.

Example

If a card is drawn at random from a pack, find the probability that it is a King.

There are fifty-two possible outcomes of which four are Kings therefore p(King) = 4/52.

The main limitation of this approach to probability is the assumption of equally likely outcomes. It could not be used, for example, if the coin was weighted in favour of heads, or in more complex problems such as the assessment of the probability of a machine breaking down.

(b) The empirical (or relative frequency) approach

This approach is based on the proportion of times an event occurs in a number of observations or experiments. For example, suppose that 1,000 light bulbs produced by a company are tested and fifty are found to be defective. The probability of a bulb being defective would be estimated to be 50/1,000 or 0.05. Thus, using the empirical approach

Of course, for this probability to remain valid, manufacturing conditions etc. must remain unchanged.

$$p(A) = \frac{\text{number of times event } A \text{ occurred}}{\text{total number of observations or experiments}}$$

Clearly, the estimate of the probability will improve as the number of observations or experiments increases.

Note that, with the empirical approach, a probability of 0 means that the event did not occur in the observed period — not that it is necessarily impossible. Similarly, a probability of 1 means that the event occurred on every possible occasion — which does not necessarily imply that it is certain to occur in the future.

(c) The subjective approach

Many problems are not suitable for the *a priori* approach, and it is often impossible to gather data which is suitable for the empirical approach. In these cases the subjective approach can be used. This involves an individual 'guestimating' the required probability. For example, a marketing manager, involved in the launch of a new product, may estimate that the probability of a break-even level of sales being achieved in the first year is about 0.7. The approach is subjective because different individuals would suggest different probabilities.

With the subjective approach, a probability of 0 means only that it is believed that the event is impossible. Similarly, a probability of 1 means that it is believed that the event is certain.

4.4 Mutually exclusive and non-mutually exclusive events

Two or more events are said to be mutually exclusive if they cannot occur together. For example, a television set cannot be both defective and non-defective at the same time. Similarly, it is not possible to draw a card which is both a King and an Ace from a pack of cards. However, Diamonds and Kings are not mutually exclusive because a selected card can be both a King and a Diamond (i.e. the King of Diamonds).

4.5 The addition rule

The addition rule is used to calculate the probability that either one event or another event will occur. Before using the rule it is important to establish whether or not the events are mutually exclusive.

p(A or B) is sometimes written as p(A UB).

If the events are mutually exclusive then

$$p(A \text{ or } B) = p(A) + p(B)$$

Example
A card is selected at random from a pack. Find the probability that it is either a King or an Ace.

$$p(\text{King or Ace}) = p(\text{King}) + p(\text{Ace}) = 4/52 + 4/52 = 8/52$$

If the events are not mutually exclusive then

$$p(A \text{ or } B) = p(A) + p(B) - p(A \text{ and } B)$$

Example
A card is selected at random from a pack. Find the probability that it is either a King or a Diamond.

$$p(\text{either King or Diamond}) = p(\text{King}) + p(\text{Diamond})$$
$$- p(\text{King and Diamond})$$
$$= 4/52 + 13/52 - 1/52 = 16/52$$

The last term, $p(\text{King and Diamond})$, is subtracted to counteract the effect of counting the King of Diamonds both as a King and a Diamond.

4.6 Complementary events

The set of outcomes *not* belonging to an event is called the complement of the event. Thus, referring to Figure 4.1, the complement of the event 'at least one head' is the event 'two tails'. The complement of event A is normally written as $\bar{A}$ and the concept leads to a useful formula:

In this context, Ā, of course, does not mean 'the mean of A'.

$$p(\bar{A}) = 1 - p(A)$$

Example
The probability that a component is defective is 0.2. Find the probability that it is not defective.

$$p(\text{not defective}) = 1 - p(\text{defective}) = 1 - 0.2 = 0.8$$

Example

The probability of no customers arriving at an accounts desk in the next hour is 0.05. What is the probability that at least one customer arrives?

$$p(\text{at least 1 customer arrives}) = 1 - p(\text{no customers arrive})$$
$$= 1 - 0.05$$
$$= 0.95$$

4.7 Independent and dependent events

Two events are independent if the occurrence of one event has no effect on the probability of the other event occurring. For example, if a coin and a die are tossed together the fact that the die gives a six will have no effect on the probability of obtaining a head on the coin. However, the announcement of disappointing profit figures by a major British company will affect the probability of share prices falling on the London Stock Exchange, so these two events are dependent.

4.8 Conditional probability

The probability of event A occurring given that event B has already occurred is called the conditional probability of event A and is denoted by $p(A|B)$.

If events A and B are independent the fact that B has occurred has no effect on the probability of A occurring so that $p(A|B) = p(A)$. Thus

$$p(\text{head on coin}|6 \text{ on die}) = p(\text{head on coin}) = 0.5$$

Conditional probability is only of interest, therefore, when the events are dependent.

Example

A light bulb is to be selected at random from a consignment of eleven bulbs, details of which are given below. Find (a) the probability that the bulb is defective; (b) the probability that the bulb is defective given that it is known to be a 60 watt bulb.

Type of bulb	Defective	Non-defective	Total
60 watts	2	5	7
100 watts	1	3	4
	—	—	—
Total	3	8	11
	—	—	—

(a) $p(\text{bulb is defective}) = 3/11 = 0.27$

(b) Because it is known that the selected bulb is 60 watts it can be only one of seven bulbs. Of these seven bulbs two are defective. Therefore

$p(\text{bulb is defective}|\text{bulb is 60 watts}) = 2/7 = 0.29$

4.9 The multiplication rule

The multiplication rule is used to calculate the probability of both one event and another event occurring. The application of the rule depends upon whether or not the events are independent.

If the events are independent then

$$p(\text{both } A \text{ and } B) = p(A) \times p(B)$$

p(A and B) is sometimes written as p(A ∩ B)

Example

A company is engaged in a civil engineering project. The probability that the completion of the project will be delayed by bad weather is 0.3 while the probability that it will be delayed by geological problems is 0.2. What is the probability that the project will be delayed by both bad weather and geological problems?

$$p(\text{bad weather and geological problems})$$
$$= p(\text{bad weather}) \times p(\text{geological problems}) = 0.3 \times 0.2 = 0.06$$

If the events are dependent then

$$p(A \text{ and } B) = p(A) \times p(B|A)$$

Example

A box contains twenty spark plugs of which six are substandard. If two plugs are selected from the box without replacement what is the probability that they are both substandard?

$$p(\text{first plug is substandard}) = 6/20$$

However, when the second plug is about to be selected only nineteen plugs remain in the box. Also, given that the first plug was substandard, only five substandard plugs remain. Thus

$$p(\text{second plug substandard}|\text{first plug substandard}) = 5/19$$

Therefore

$$p(\text{both plugs substandard})$$
$$= p(\text{first substandard}) \times p(\text{second substandard}|\text{first substandard})$$
$$= 6/20 \times 5/19 = 30/380 = 0.079$$

4.10 Probability trees

Probability trees are useful for clarifying awkward problems.

Example

The probability of a person seeing a television advertisement for a product is 0.3. If the person sees the advertisement the probability that the product will be purchased is 0.4. However, if the advertisement has not been seen, the probability that a purchase will be made is only 0.1. What is the probability the person will purchase the product?

The probability tree in Figure 4.2 shows all the combinations of events

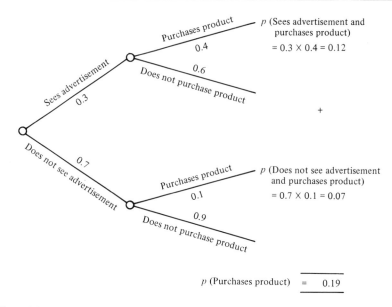

Figure 4.2

The formula given for complementary events has been used here. For example, p (sees advertisement) = 0.3. Therefore p (did not see advertisement) = 0.7.

which can occur. Note that whether or not the person saw the advertisement comes at the start of the tree since this influences his or her chances of subsequently purchasing the product. Both the multiplication and addition rules have been used to determine the final probability, which is 0.19.

4.11 Bayes theorem

The term, prior probability, should not be confused with the a priori approach to probability.

Bayes theorem enables an initial estimate of a probability, known as a *prior probability*, to be modified in the light of new information. The result is called a *posterior probability*. For example, a manager's initial estimate of the probability of a product having high sales in the coming year may be revised upwards when favourable market research results become available.

The formula for Bayes theorem is

$$p(A|B) = \frac{p(A) \times p(B|A)}{p(B)}$$

This is a simple way of stating Bayes theorem. You may see it stated in other ways in some textbooks.

where $p(A)$ is the prior probability of A occurring and B represents the new information which has been obtained. $p(B)$ can be obtained from a probability tree.

Example
A manager estimates that there is a 0.7 probability that a production process is running at peak efficiency. However, he is then shown the latest production cost figures which have exceeded the budgeted figure. The probability of the costs exceeding the budgeted figure if the process is at

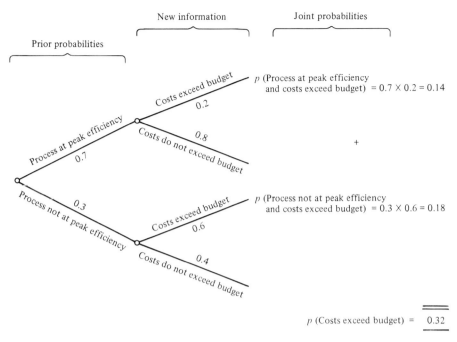

Figure 4.3

peak efficiency is only 0.2, while the probability of them exceeding the budgeted figure if the process is not at peak efficiency is 0.6. Revise the manager's estimate in the light of the new information.

$$p(A) = \text{prior probability of process being at peak efficiency}$$
$$= 0.7$$
$$p(B|A) = p(\text{costs exceed budget} | \text{process at peak efficiency})$$
$$= 0.2$$
$$p(B) = p(\text{costs exceed budget})$$
$$= 0.32 \text{ (from the tree in Figure 4.3)}$$

Therefore from the formula

$$p(\text{process at peak efficiency} | \text{costs exceed budget})$$

$$= \frac{0.7 \times 0.2}{0.32} = \frac{0.14}{0.32} = 0.438$$

Note that, on the tree, the first branches are used for the prior probabilities and the later branches relate to the information which has been obtained.

Thus the manager's initial estimate that the process is at peak efficiency should be revised from 0.7 to 0.438 in the light of the cost information.

4.12 Permutations

Suppose that a travelling salesman is to visit three towns: *A*, *B* and *C*. He can choose any of three towns for his first visit, then either of the two remaining towns for the second visit and his last visit will then be predetermined. The number of possible routes is therefore $3 \times 2 \times 1 = 6$ (i.e. *ABC, ACB, BAC, BCA, CAB, CBA*). This product can be written as 3! (3 factorial). In general *n* different objects can be ordered or arranged in *n*! ways where

$$n! = n \times (n-1) \times (n-2) \times \ldots \ldots \times 2 \times 1$$

Thus $5! = 5 \times 4 \times 3 \times 2 \times 1 = 120$.

Note that $0! = 1$ and $1! = 1$. Each possible arrangement is called a permutation.

Often it is necessary to calculate the number of possible permutations which exist when just a few objects are selected from a population of objects. For example, suppose that from a shortlist of four people, *W, X, Y* and *Z*, a sales clerk and then a wages clerk are to be selected. Any of four people could be chosen for the sales clerk job which leaves any three for the wages job. There are thus $4 \times 3 = 12$ possible permutations of appointments. In general if *r* objects are selected from a population of *n* objects the number of possible permutations is given by ^nP_r where

$$^nP_r = \frac{n!}{(n-r)!}$$

In this case, $n = 4, r = 2$ so that

$$^4P_2 = \frac{4!}{(4-2)!} = \frac{4 \times 3 \times 2 \times 1}{2 \times 1} = 12$$

4.13 Combinations

With permutations the order of selection is important. Thus, in the above example, the first appointment is the sales clerk and the second the wages clerk so that *WX* is a different permutation to *XW*. However, with combinations the *order does not matter*. Suppose this time two delegates are to be selected from *W, X, Y* and *Z* for a conference. Clearly, there is nothing to distinguish the selection *WX* from *XW* because the same two people will go to the conference. These two selections therefore represent the same combination. Generally, if *r* objects are to be selected from a population of *n* objects the number of possible combinations is given by ^nC_r where

$$^nC_r = \frac{n!}{r!(n-r)!}$$

For the delegates example, $n = 4$ and $r = 2$

$$\text{so } ^4C_2 = \frac{4!}{2!(4-2)!} = \frac{4 \times 3 \times 2 \times 1}{2 \times 1 \times 2 \times 1}$$

$$= 6 \text{ possible combinations}$$

These are *WX, WY, WZ, XY, XZ* and *YZ*.

4.14 Probability distributions

A statement of all the possible values a variable can assume, together with the probability of it assuming these values, is called a probability distribution. For example, the probability distribution of the weekly sales of a product is given below.

Sales (no. of units)	0	1	2	3	4	Total
Probability	0.1	0.3	0.4	0.1	0.1	1.0

Note that the probabilities sum to one because all the possible levels of sales have been given and it is therefore certain that one or other of the sales levels will occur.

4.15 Expected values

The mean of a probability distribution is called an expected value, and it can be interpreted as a 'long-run average result'. The calculation of an expected value simply involves (a) multiplying each value by its probability of occurrence and (b) adding the results. Thus for the probability distribution of sales given in the last section

Note the difference between the use of the word 'expected' in its statistical sense (meaning a long-run average) and the use of the word in everyday language.

$$\text{Expected sales} = (0 \times 0.1) + (1 \times 0.3) + (2 \times 0.4) + (3 \times 0.1)$$
$$+ (4 \times 0.1)$$
$$= 0 + 0.3 + 0.8 + 0.3 + 0.4 = 1.8 \text{ units}$$

This means that over a large number of weeks the mean level of sales should be about 1.8 units per week.

4.16 Decision analysis

This topic does not appear on the CACA(1.5) syllabus

4.16.1 Decision trees

A decision tree is a diagrammatic representation of a decision problem. It consists of two symbols. A square represents a decision box. Immediately beyond this the decision maker can decide which branch to follow. A circle is a 'chance node' and immediately beyond this chance determines which branch is followed.

Example
An important machine in a factory has developed a fault and the engineer has to decide whether to return it to the supplier (at a cost of £8,000) or to attempt to repair it himself (at an estimated cost of £3,000). The probability that he will fail to repair the machine himself is 0.5. If he does fail, he will then have to decide whether to make a second attempt at repair (by investigating a different part of the mechanism) at an additional cost of £3,000, or to return the machine to the supplier after all. The probability that he will fail at the second attempt is 0.7 and, if he does fail, the machine

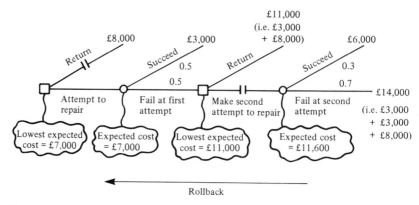

Figure 4.4

Note that, on a decision tree, probabilities
should only be written on branches
immediately following a circle, otherwise it
would imply that the decision maker is
leaving his or her decisions to chance.

will definitely have to be returned to the supplier. Draw a decision tree and
recommend a course of action.

Figure 4.4 shows the decision tree for this problem. The decision which is
made depends upon the decision criterion which is adopted. The widely
used expected monetary value (EMV) criterion implies that the best
course of action is the one which yields the highest expected return or the
lowest expected cost.

In order to find the best sequence of decisions a procedure known as the
'rollback technique' is adopted. This involves considering the later
decisions first. Thus, assuming for the moment that the first attempt at
repair has failed, should the engineer make a second attempt at repair or
return the machine?

Expected cost of second attempt
$$= 0.3 \times 6,000 + 0.7 \times 14,000 = £11,600$$
while the expected cost of returning the machine $\quad = £11,000$

Therefore, if the first attempt fails, the machine should be returned. The
inferior course of action is now crossed out on the tree and the expected
cost of the best course of action (£11,000) is written next to the decision
box. It is then treated as the cost at the end of the 'fails at first attempt'
branch.

Moving back to the initial decision:

Expected cost of returning machine at outset = £8,000
Expected cost of first attempt at repair
$$= 0.5 \times 3,000 + 0.5 \times £11,000 = £7,000$$

Therefore, initially, the engineer should attempt to repair the machine
himself. If he fails, he should return the machine rather than making a
second attempt.

The method used above has a number of limitations. The probabilities and
costs used are likely to be only rough estimates. This could be important,
given the closeness of some of the expected values. Also, the use of

This does not mean that, if the best
alternatives are chosen, costs of £7,000 will
be incurred. The engineer will either incur
costs of £3,000 or £11,000. The £7,000 is
simply the average cost which he would
incur if he made the decision a large
number of times.

expected values (which predict average costs if the decision is made a large number of times) is questionable if this is a one-off decision.

The remainder of this chapter is only relevant to the ACCA(2.6) and CIPFA syllabuses

4.16.2 Decision tables

Decision tables can also be used to represent decision problems. For example, suppose that a retailer must decide whether to hold small or large stocks of a commodity for the coming month. His or her profit for the month will depend on the level of demand for the commodity as shown in the decision table below:

(Profits)		Outcomes 0,4	0 6
		Low demand	High demand
Courses	Hold small stocks	£4,000	£7,000
of action	Hold large stocks	£1,000	£9,500

If the probabilities of low and high demand are estimated to be 0.4 and 0.6 respectively then:

Expected profit of holding small stocks
$$= 0.4 \,(£4,000) + 0.6 \,(£7,000) = £5,800$$

and expected profit of holding large stocks
$$= 0.4 \,(£1,000) + 0.6 \,(£9,500) = £6,100$$

Thus, according to the EMV criterion, large stocks should be held.

4.16.3 Opportunity loss

Opportunity loss is the cost of not having made the best decision when a particular outcome occurs. For example, if demand turned out to be low and large stocks were held, profit is £3,000 lower than it could have been. Opportunity loss can thus be found by subtracting every payoff in a column of the decision table from the best payoff in the column. The opportunity loss table for the retailer's problem is therefore:

		Outcomes	
		Low demand	High demand
Courses	Hold small stocks	£0	£2,500
of action	Hold large stocks	£3,000	£0

If the retailer wishes to minimize expected opportunity loss (EOL) then:

$$\text{EOL of holding small stocks} = 0.4\,(£0) + 0.6\,(£2{,}500) = £1{,}500$$
$$\text{and EOL of holding large stocks} = 0.4\,(£3{,}000) + 0.6\,(\quad £0) = £1{,}200$$

which again suggests that large stocks should be held. In fact, the EMV and EOL criteria always lead to the same decision.

4.16.4 Expected value of perfect information (EVPI)

The expected value of perfect information is the expected increase in payoff which would result if uncertainty could be removed from the decision. If the retailer was able to find out, at the start of each month, what the level of demand was going to be then he would always make the best decision. If demand was going to be low he would hold small stocks (thus earning £4,000 profit). Similarly, if demand was going to be high he would hold large stocks (earning £9,500). The probabilities imply that low demand occurs in 40 per cent of months and high demand in 60 per cent thus:

Expected (or average) profit *with* perfect information
$$= 0.4\,(£4{,}000) + 0.6\,(£9{,}500) = £7{,}300$$
The best expected profit *without* perfect information $\qquad = \underline{£6{,}100}$
Therefore: expected value of perfect information $\qquad = £1{,}200$

Thus the retailer's profit would increase by an average of £1,200 per month if he could obtain perfect information. Note that the EVPI always equals the minimum EOL.

4.16.5 Expected value of imperfect information

Information used by decision makers often comes from surveys based on samples, and it is not, therefore, perfectly accurate. Suppose that the retailer intends to conduct market research which will predict either high or low demand for the month. The reliability of this research is shown by the following probabilities:
   If demand in the month will be low:

p(research predicts low demand) $= 0.8$
p(research predicts high demand) $= 0.2$

   If demand in the month will be high:

p(research predicts low demand) $= 0.4$
p(research predicts high demand) $= 0.6$

   The expected value of this imperfect information can be calculated as follows:
(i)   Calculate the probability that research will predict low demand.
   The probability tree in Figure 4.5(a) shows that this probability is 0.56.
(ii)   Use Bayes theorem to modify the probabilities of low and high demand, given a prediction of low demand.

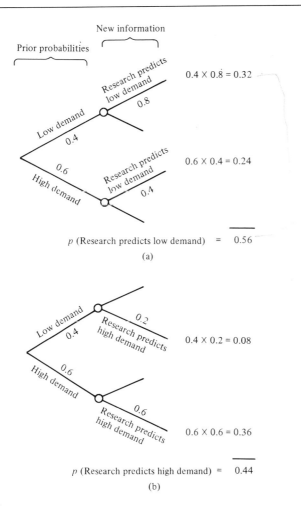

Figure 4.5

From Figure 4.5(a) it can be seen that the posterior probability of low demand is 0.32/0.56 which is 0.57 and the posterior probability of high demand is 0.24/0.56 which is 0.43.

(iii)   Identify the optimal decision, given these posterior probabilities.

Applying the posterior probabilities to the retailer's decision table:

The expected profit of holding small stocks = £5,290
The expected profit of holding large stocks = £4,655

Thus, if market research predicts low demand, small stocks should be held.

(iv)   repeat the above steps for a prediction of high demand.

Figure 4.5(b) shows that p(research predicts high demand) = 0.44 and the posterior probabilities of low and high demand are 0.18 (i.e. 0.08/0.44) and 0.82 (i.e. 0.36/0.44) respectively. Applying these probabilities to the decision table it can be seen that the expected profit of holding small stocks is £6,460, while the expected profit of holding large stocks is £7,970 so large stocks should be held if high demand is predicted.

Thus the expected profit *with* the imperfect information
= p(low demand predicted)
    × £5,290 + p(high demand predicted) × £7,970
                    = 0.56 (£5,290) + 0.44 (£7,970) = £6,469.2
Without the imperfect information the expected profit   = £6,100
                                                             ───────
Therefore: expected value of imperfect information   =   £369.2

4.16.6 Utility theory

The EMV and EOL criteria may not be appropriate for all decision makers. For example, given a choice of receiving £3,000 for certain or participating in a gamble which offered a 50 per cent chance of losing £2,000 and a 50 per cent chance of gaining £10,000, many people would opt for the certain money, even though the expected payoff of the gamble is £4,000. Utilities can be used to represent such preferences. Note that U(£8,000) means 'the utility of £8,000'.

4.16.7 Measuring utility

One method of measuring utility involves assigning the best possible payoff of a decision a utility of 100 and the worst possible payoff, a utility of 0. Thus, for the retailer, U(£9,500) = 100 and the U(£1,000) = 0. The utility of other payoffs can then be found by offering the decision maker a choice between receiving these payoffs for certain or entering a gamble which will result in either the best or worst payoff. The probabilities in the gamble are varied until the decision maker is indifferent between the certain money and the gamble.

Suppose that the retailer is indifferent between receiving £7,000 for certain and a gamble offering a 0.9 probability of £9,500 (the best payoff) and a 0.1 probability of £1,000 (the worst payoff). Then the utility of £7,000 equals the expected utility of the gamble:

i.e. U(£7,000) = 0.9 U(£9,500) + 0.1 U(£1,000) = 0.9 (100) + 0.1 (0) = 90.

U(£4,000) could be similarly determined. Suppose this is 50. The utilities would then replace the monetary values in the decision table and the course of action yielding the highest expected utility could then be identified. For the retailer:

   expected utility of small stocks = 0.4 (50) + 0.6 (90) = 74
and expected utility of large stocks = 0.4 (0) + 0.6 (100) = 60

so the retailer should hold small stocks.

5
Probability distributions

The material in this chapter appears on the CIMA, ICAEW, ICAI, ICAS, CACA(2.6), CIPFA and ICSA syllabuses

5.1 Introduction

Some probability distributions have well-defined patterns which can be represented by mathematical formulae. Three such distributions, the binomial, Poisson and normal distributions, are considered next. Though they have many applications, these distributions can be particularly useful when data which has been obtained from random samples needs to be interpreted.

5.2 The binomial distribution

This topic does not appear on the ICSA syllabus

Consider the following problem. A salesman has five prospective customers to visit and wants to know the probability that four of the customers will place orders. If each visit to a customer is regarded as a 'trial' which can have two possible outcomes, 'success' (i.e. an order is secured) or 'failure', then the salesman's problem can be expressed in general terms as

$$p(\text{obtaining } r \text{ successes in } n \text{ trials})$$

where, in his case, $r = 4$ and $n = 5$.

Other examples of this type of problem might be p(finding 2 defective light bulbs in a sample of 10 bulbs) or p(10 people respond to a survey if 50 questionnaires are sent out).

If certain conditions are fulfilled, the binomial distribution can be used to calculate probabilities like these.

5.2.1 Conditions necessary for using the binomial distribution

1 The trials must be independent. For the above salesman, the fact that one customer places an order should in no way affect the chances of the other customers placing orders.
2 The probability of a success must be the same for each trial. The salesman must have the same probability of securing an order from each customer.

The word 'success' does not imply that the occurrence of the event is desirable. For example, a 'success' may be the occurrence of a defective item in a sample or a machine breaking down.

Sometimes the binomial distribution can be applied approximately when p is not the same for each trial. For example, suppose that television sets are sampled from a warehouse containing 1000 sets of which 100 are defective. The probability of the first set selected being defective is 100/1000 i.e. 0.1. If this set is not replaced the probability of the second set being defective is 99/999 i.e. 0.099 and so on. As a rule of thumb these changes in the value of p will be insignificant if the size of the sample is less than 20 per cent of the size of the population.

3   The trials must have only two possible outcomes (i.e. success and failure). If the salesman wished to calculate the probability that in his five visits two customers will place 'large orders', two will place 'small orders' and one 'will not place an order' then the binomial distribution will not apply because each trial can have three possible outcomes.

5.2.2 Calculating probabilities using the binomial distribution

The formula for calculating probabilities using the binomial distribution is

$$p(r \text{ successes in } n \text{ trials}) = {}^{n}C_{r}p^{r}(1-p)^{n-r}$$

where p = the probability of a success occurring in a single trial.

Thus to use the formula three values need to be specified: n, the number of trials, r, the number of successes for which the calculation is being carried out and p.

Example

The salesman referred to above estimates that the probability of securing an order from a customer is 0.4. What is the probability that four out of the five customers will place orders?

For this problem: $n = 5$, $r = 4$ and $p = 0.4$. Therefore

$$p(4 \text{ customers place orders out of the 5 visited})$$
$$= {}^{5}C_{4}(0.4)^{4}(1-0.4)^{5-4} = 5(0.0256)(0.6)$$
$$= 0.0768$$

Example

Find the probability that the above salesman secures orders from (a) four or more customers, (b) less than four customers.

(a)   four or more includes the possibility of either four or five customers placing orders. Using the addition rule of probability

$$p(4 \text{ or } 5 \text{ place orders}) = p(4 \text{ place orders}) + p(5 \text{ place orders})$$
$$= 0.0768 + {}^{5}C_{5}(0.4)^{5}(1-0.4)^{5-5}$$
$$= 0.0870$$

(b)   Less than four includes the possibility of zero, one, two or three customers placing orders. However, the quickest way to calculate this probability is to note that less than four is the complement of four or more. Therefore

$$p(\text{less than 4 place orders}) = 1 - p(4 \text{ or more place orders})$$
$$= 1 - 0.0870 \text{ (from above)}$$
$$= 0.9130$$

5.2.3 The mean and standard deviation of the binomial distribution

For the binomial distribution

Mean (or expected) number of successes in n trials $= np$
Standard deviation $= \sqrt{np(1-p)}$

Thus if the salesman visits five customers each day the mean number of customers placing an order will be $5 \times 0.4 = 2$ and the standard deviation of the number of customers placing orders will be $\sqrt{5 \times 0.4 \times 0.6} = 1.1$ customers. The standard deviation will reflect the extent to which the number of customers placing orders varies from day to day.

5.3 The Poisson distribution

This topic does not appear on the ICSA syllabus

The Poisson distribution can be used to calculate the probability of an event occurring a specified number of times over a certain period of time or within a particular length, area or volume of space. For example, a company maintenance department may be interested in knowing the probability of three machine break-downs occurring in the next week. A carpet manufacturer may wish to know the probability of two flaws in the pattern occurring in a 90 metre length of carpet.

5.3.1 Conditions necessary for using the Poisson distribution

1   The probability of the event occurring must be the same for each small interval of time or unit of space. For example, there must be the same probability of a machine break-down occurring between 11.04 and 11.05 on a Wednesday morning as between 3.14 and 3.15 on a Friday afternoon.
2   The events must occur independently. The fact that one machine has just broken down should not increase or reduce the probability of another machine break-down occurring.
3   Events cannot occur together at exactly the same time.

5.3.2 Calculating probabilities using the Poisson distribution

The formula for the Poisson distribution is

$$p(r \text{ occurrences of an event}) = \frac{e^{-m}m^r}{r!}$$

where e is the number 2.71828 . . . and m is the mean number of occurrences for the period of time (or space) under consideration. Normally a calculator with an e^x button can be used to evaluate e^{-m}. It can be seen that only m and r need to be specified to calculate a Poisson probability. (CIMA students can obtain Poisson probabilities directly by using Table 9 in *Mathematical Tables for Students*.)

This apparently rather obscure number has some very important mathematical properties and, like π, cannot be stated exactly in a finite number of decimal places.

Example
Arrivals at a car exhaust fitting centre follow a Poisson distribution with a mean of two per hour.

(a)   Find the probability of (i) no (ii) one and (iii) two or more customers arriving in the next hour.

(b)   Find the probability of three customers arriving in the next two hours.

(a)   (i) $m = 2$, $r = 0$ so that $p(0 \text{ arrivals}) = e^{-2}\,2^0/0! = 0.1353$
     (ii) $m = 2$, $r = 1$ so that $p(1 \text{ arrival}) = e^{-2}\,2^1/1! = 0.2707$
     (iii) $p(2 \text{ or more arrivals}) = 1 - p(0 \text{ or } 1 \text{ arrival})$
$$= 1 - (0.1353 + 0.2707) = 0.5940$$

(b)   Since the mean arrival rate is two per hour it will be four for each two-hour period. Therefore $m = 4$ and $r = 3$. Thus $p(3 \text{ arrivals in two hours}) = e^{-4}\,4^3/3! = 0.1954$.

The standard deviation of a Poisson distribution is equal to the square root of its mean. For the above example, the standard deviation of the number of customers arriving per hour will be $\sqrt{2}$ that is 1.414 customers. This measures the extent to which the number of customers arriving varies from hour to hour.

5.3.3 The Poisson approximation to the binomial distribution

Calculating probabilities using the binomial distribution can be very tedious when the number of trials (n) is large. However, the Poisson distribution gives a good approximation to the binomial when: (i) n is at least 30 *and* (ii) p is less than 0.1.

Example
One per cent of a company's customers default on payments. If the payments of 300 randomly sampled customers are monitored, find the probability that two of these customers will default on payment.
For this problem $n = 300$, $r = 2$ and $p = 0.01$. The binomial formula requires a rather awkward expression to be evaluated since

$$p(2 \text{ customers default out of } 300) = {}^{300}C_2\,(0.01)^2\,(0.99)^{298}$$

To use the Poisson approximation m is set equal to np. Thus $m = 300 \times 0.01 = 3$. Therefore

$$p(2 \text{ customers default}) = e^{-3}\,3^2/2! = 0.2240$$

5.4 The normal distribution

Consider Figure 5.1 which shows a histogram of the weights of bags of flour which have been filled by a machine. It can be seen that the distribution can be approximately represented by a bell-shaped, symmetrical curve.

Many variables have distributions of this shape and are said to be normally distributed. The exact shape of a normal distribution is determined by its mean (μ) and standard deviation (σ).

The binomial and Poisson distributions are examples of *discrete distributions* since they represent situations where the number of occurrences of an event (e.g. the number of defectives discovered or number of customers arriving) must be a whole number. However, the normal distribution is a *continuous distribution* since it represents variables which

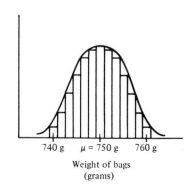

740 g $\mu = 750\text{ g}$ 760 g
Weight of bags
(grams)

Figure 5.1

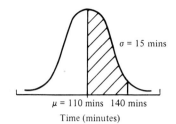

Figure 5.2

can theoretically take on any value within the range of possible values. Nevertheless, the probability of a bag of flour, for example, weighing exactly 754.231 grams is very, very small. It is therefore only meaningful to calculate the probability of the variable taking on a value within a given range, for example between 750 and 760 grams.

5.4.1 Areas under the normal curve

Suppose that a company producing a large number of electronic components has found that the time it takes a worker to assemble a component is normally distributed with a mean of 110 minutes and a standard deviation of 15 minutes. The production manager wants to know the probability that a component which has just been ordered will take between 110 and 140 minutes to assemble.

It is always a good idea to draw a diagram when solving normal distribution problems.

Figure 5.2 illustrates the problem. The required probability is represented by the area under the curve which has been shaded (the total area under the curve is 1). To determine this area use is made of the Table of areas under the normal curve (Appendix 1). Note that the required area is an area between the mean and another value and the tables are designed to give this sort of area directly. To use the tables, 140 minutes has to be converted to a Z value where

This relates to the fact that for a histogram area is proportional to frequency.

$$Z = \frac{x - \mu}{\sigma} = \frac{\text{the original value} - \text{mean}}{\text{standard deviation}}$$

$$= \frac{140 - 110}{15} = 2$$

Note that the tables in Appendix 1 are those issued to CIMA students. Some professional bodies (e.g. the CACA) use tables which give the 'area in the tail' of the normal distribution (i.e. the area to the right of a point, rather than the area between the mean and the point).

A Z value gives the number of standard deviations of a point from the mean. Thus a component which takes 140 minutes to assemble takes two standard deviations longer than the mean assembly time. From the tables it can be seen that the required area, and hence the required probability, is 0.4772.

Example

For the problem described above, calculate the probability that the assembly time of the component will be (a) between 80 and 110 minutes; (b) over 140 minutes; (c) between 125 and 140 minutes.

Referring to Figure 5.3(a) to (c)

(a)   For 80 minutes

$$Z = \frac{80 - 110}{15} = -2$$

However, because of the symmetry of the distribution, the area between 80 minutes and 110 minutes is the same as the area between 110 and 140 minutes, so the answer is 0.4772 as before. Therefore, when using the tables, the negative sign of the Z value can be ignored.

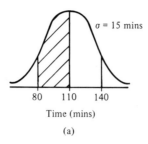

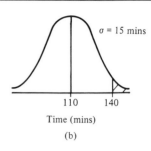

Time (mins)

(a)

Time (mins)

(b)

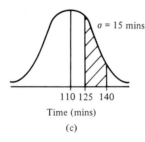

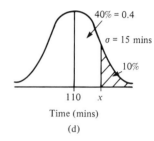

Time (mins)

(c)

Time (mins)

(d)

Figure 5.3

(b)   Half of the area under the curve lies above the mean. Therefore

Area above 140 minutes $= 0.5 -$ area between 110 and 140 minutes
$$= 0.5 - 0.4772 = 0.0228$$

(c)

Required area $=$ area between 110 and 140 minutes
$- $ area between 110 and 125 minutes

For 125 minutes

$$Z \text{ value} = \frac{125 - 110}{15} = 1$$

so the required area $= 0.4772 - 0.3413 = 0.1359$.

Example
For the distribution given above determine the assembly time which is
exceeded by 10 per cent of components.
   Figure 5.3(d) illustrates the problem, which is the opposite of the earlier
problems. Here the area is given, but it is necessary to determine the
assembly time (x). If 10 per cent of components have an assembly time
greater than x, then 40 per cent must have an assembly time between 110
minutes and x. 40 per cent is converted to a decimal i.e. 0.4 and the 'main

body' of the table is searched until a value as close as possible to 0.4 is found. The closest value is 0.3997 and this corresponds to a Z value of 1.28. This means that x is about 1.28 standard deviations from the mean. The standard deviation is 15 minutes and Figure 5.3(d) shows that x is above (rather than below) the mean therefore

$$x = 110 + 1.28(15) = 129.2 \text{ minutes}$$

Thus 10 per cent of components will take longer than about 129 minutes to assemble.

5.4.2 Adding normal distributions

Some problems require normal distributions to be added. Consider the following problem. A lorry regularly delivers goods to a customer. The time it takes to travel to the customer is normally distributed with a mean of 8 hours and a standard deviation of 2 hours. The return journey time is also normally distributed with a mean of 6 hours and a standard deviation of 1.5 hours and is independent of the time for the outward journey. What is the probability that the total travelling time will exceed 18 hours?

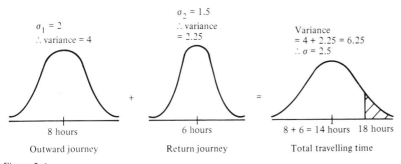

Figure 5.4

Figure 5.4 illustrates the problem which is solved as follows.

The mean total travelling time is $8 + 6 = 14$ hours. Because the times for the journeys are independent the total time will also be normally distributed. Also, because the distributions are independent, their *variances* can be added. Thus

$$\begin{aligned}
\text{Variance of total time} &= \text{variance of outward journey} \\
&\quad + \text{variance of return journey} \\
&= 2^2 + (1.5)^2 = 6.25
\end{aligned}$$

Therefore the standard deviation of the total time $= \sqrt{6.25} = 2.5$. Eighteen hours has a Z value of

$$\frac{18 - 14}{2.5} = 1.6$$

so the required probability is $0.5 - 0.4452 = 0.0548$.

5.4.3 The normal approximation to the binomial distribution

In Section 5.3.3 the Poisson distribution was used to approximate the binomial distribution when n, the number of trials, was large. However, this approximation was only valid when p was less than about 0.1. If this is not the case, an alternative solution is to use the normal distribution as an approximation to the binomial. This approximation will be reasonable if np is greater than five and p is between 0.1 and 0.9.

To use the approximation the mean of the normal distribution is set equal to np and the standard deviation to $\sqrt{np(1-p)}$. However, because the binomial is a discrete distribution, a continuity correction has to be used. This reflects the fact that, with the normal distribution, one can only

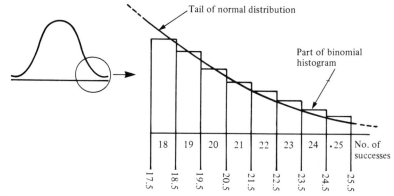

Figure 5.5

think in terms of a variable having a value within a given range rather than an exact value. Thus the probability of twenty successes will be represented by the area under the normal curve between 19.5 and 20.5 while the probability of more than twenty-four successes will be regarded as the area to the right of 24.5. Figure 5.5 illustrates this.

Example

Twenty-five per cent of orders received by a company qualify for bulk-purchase discounts. If a random sample of 200 orders is selected what is the probability that sixty or more of the orders will be eligible for the discount? Here $n = 200$, $p = 0.25$. Therefore

$$\mu = np = 200 \times 0.25 = 50$$
$$\text{and } \sigma = \sqrt{np(1-p)} = \sqrt{200(0.25)(0.75)} = 6.12$$

Figure 5.6 illustrates the normal curve for this problem. The probability of sixty or more orders is represented by the area to the right of 59.5 (because sixty or more includes sixty). For 59.5 the Z score is

$$\frac{59.5 - 50}{6.12} = 1.55$$

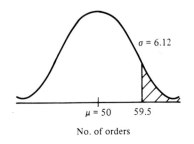

Figure 5.6

The tables show that the area between the mean and 59.5 is 0.4394. Therefore the required probability is $0.5 - 0.4394 = 0.0606$.

6
Estimation and hypothesis testing

The material in this chapter is relevant to the following syllabuses, except where otherwise indicated, CIMA, ICAEW, ICAI, ICAS, CACA(2.6), CIPFA, ICSA

6.1 Introduction

In Chapter 1 the use of samples to obtain statistical information was discussed. This information is generally used to provide an estimate of statistics relating to the population, or to test a hypothesis about the population. This chapter describes how estimation and hypothesis testing can be carried out when a simple random sample has been selected from the population.

6.2 Estimation

Two statistics which often need to be estimated from samples are

1 *The mean of the population.* For example, in quality control it may be necessary to estimate, on the basis of a sample, the mean weight of packets of frozen food leaving a production line.
2 *The percentage of the population having a given characteristic.* For example, a mail-order company may wish to estimate, by monitoring a sample of orders, the percentage of all orders which take longer than fourteen days to deliver.

6.2.1 Point estimates and confidence intervals

An estimate derived from a sample which is expressed as a single figure (e.g. 'we estimate that the mean weight of our packets of frozen food is 450 grams') and is known as a *point estimate*.

However, because the estimate is based on only a sample of items, it is often more desirable to give a range of figures which is likely to contain the true value (e.g. 'we estimate that the mean weight is between 447 and 453 grams'). This is known as an *interval estimate*.

If the probability of the interval containing the true value is stated (e.g. 'there is a 95 per cent probability that the interval 447 to 453 grams includes the mean weight of the population') then the result is called a *confidence interval*.

6.2.2 Confidence intervals for the population mean based on the normal distribution

The normal distribution can be used to derive a confidence interval for the population mean when either (i) the sample size is at least 30 or (ii) the population is normally distributed *and* the population standard deviation is known.

   If either of these conditions is satisfied (and provided that the population is large relative to the sample size) the formula for a 95 per cent confidence interval for the population mean is:

$$\bar{x} \pm 1.96 \frac{\sigma}{\sqrt{n}}$$ where $\bar{x}$ = the sample mean, n = the sample size and σ = the population standard deviation

If the sample size is at least 30, the sample standard deviation, s, can be used as an approximation for σ. For a 99 per cent confidence interval 2.58 is substituted for 1.96.

6.2.3 Derivation of the confidence interval formula

Suppose that a random sample of fifty jars of coffee is taken from a production line and the mean weight of the fifty jars is calculated. The

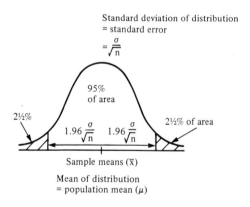

Figure 6.1   The sampling distribution of means

result is known as a *sample mean*. If the process of taking a sample of fifty jars is repeated a large number of times, a large number of sample means will be generated. If these are then grouped into a frequency distribution, the result is known as the *sampling distribution of means* and it can be shown that this will have three characteristics (see Figure 6.1).
1   It will be approximately a normal distribution.
2   The mean of the distribution will be the population mean which suggests that there is a tendency for random samples to have means which cluster around the population mean.

3   The standard deviation of the distribution, which is generally known as
     the standard error of means, will be equal to

$$\frac{\text{Population standard deviation}}{\sqrt{\text{sample size}}} = \frac{\sigma}{\sqrt{n}}$$

The formula shows that the larger the sample is then the smaller the
standard error will be. If the standard error is small it is likely that an
individual sample mean will be close to the population mean.

The normal curve tables reveal that 95 per cent of the area under a
normal curve lies within 1.96 standard deviations of the mean. Now the
sampling distribution of means is normal with a mean equal to the
population mean and a standard deviation called the standard error.
Therefore there is a 95 per cent probability that a sample mean ($\bar{x}$) will fall
within 1.96 standard errors of the population mean. Thus there is a 95 per
cent probability that the following interval will include the population
mean

$$\bar{x} \pm 1.96\sigma/\sqrt{n}$$

6.2.4 Calculating confidence intervals for the population mean

Example

A random sample of fifty invoices is taken from a company's file. The mean
value of the invoices sampled is £168.90 and the standard deviation is
£23.60. Derive a 95 per cent confidence interval for the mean value of all
the company's invoices.

$\bar{x}$ = £168.90, s = £23.60 and n = 50, therefore the 95 per cent
confidence interval is

$$168.90 \pm 1.96\frac{(23.60)}{\sqrt{50}}$$

$$= £168.90 \pm £6.54 \text{ or } £162.36 \text{ to } £175.44$$

There is a 95 per cent probability that this interval includes the mean value
of all the company's invoices.

6.2.5 Confidence intervals for percentages

When the percentage of the population with a given characteristic (i.e. the
population percentage) needs to be estimated a similar argument to that
developed in the preceding section applies. The percentage of items in the
sample with the given characteristic is known as the sample percentage
(normally represented by p). As long as the sample is large enough, the
distribution of sample percentages is approximately normal with a mean

*Confidence intervals are very often
misinterpreted. The interval here does not
mean that 95 per cent of invoices are for
amounts between £162.36 and £175.44. It
means that there is a 95 per cent
probability that the mean value of all the
company's invoices is between £162.36 and
£175.44.
Note that the cost of being more confident of
including the population mean in the
interval is a wider, or less precise, interval.*

equal to the population percentage, and a standard deviation known as the standard error of percentages where

$$\text{Standard error of percentages} = \sqrt{\frac{p(100-p)}{n}} \text{ (where } n = \text{sample size)}$$

The formula for a 95 per cent confidence interval for the population percentage is

$$p \pm 1.96 \sqrt{\frac{p(100-p)}{n}}$$

For a 99 per cent interval 2.58 is substituted for 1.96 as before.

Example
A manufacturing company wishes to estimate the percentage of electrical components produced on a particular day which are defective. In a random sample of 200 components thirty are found to be defective. Estimate, at the 95 per cent level of confidence the percentage of all the components produced which are defective.

$p = 30/200 = 15$ per cent, $n = 200$. Therefore the 95 per cent confidence interval is

$$15 \pm 1.96 \sqrt{\frac{15(100-15)}{200}}$$

$$= 15\% \pm 4.95\% \text{ or } 10.05\% \text{ to } 19.95\%$$

It may be that this interval is too wide to be useful. A larger sample would, of course, give a narrower interval.

Thus the company can be 95 per cent confident that between 10.05 per cent and 19.95 per cent of the components were defective.

6.2.6 Confidence intervals for the population mean based on the t distribution

This topic does not appear on the CIMA, ICAEW, ICAS or ICSA syllabuses

The t distribution is used to calculate confidence intervals for the population mean when:
1   The sample size is less than 30
2   The population standard deviation is not known
3   The population is at least approximately normally distributed.
For a 95 per cent confidence interval the formula is:

$$\bar{x} \pm t_{0.025} \frac{s}{\sqrt{n}}$$

where $\bar{x}$ = the sample mean, n = the sample size and s = the sample standard deviation. $t_{0.025}$ is found from Table 6.1 where the number of degrees of freedom = the sample size − 1.

The 0.025 value is used because, if the middle 95 per cent of the area under a t distribution is shaded, 2.5 per cent (i.e. 0.025) of the area will fall in each tail. Thus for a 99 per cent interval $t_{0.005}$ is used.

Table 6.1   Extract from t tables

	Number of degrees of freedom				
	1	2	3	4	5
$t_{0.050}$	6.31	2.92	2.35	2.13	2.02
$t_{0.025}$	12.71	4.30	3.18	2.78	2.57
$t_{0.010}$	31.82	6.97	4.54	3.75	3.37
$t_{0.005}$	63.66	9.93	5.84	4.60	4.03

Example

The mean distance travelled last month by four randomly selected salespeople, who are employed by a large company, was 2,100 miles with a standard deviation of 300 miles. Estimate the mean distance travelled by all the company's salespeople at the 95 per cent level of confidence.

$\bar{x}$ = 2,100 miles, s = 300 miles, n = 4 so there are three degrees of freedom. Therefore the 95 per cent confidence interval

$$= 2,100 \pm 3.18 \frac{(300)}{\sqrt{4}}$$

$$= 1,623 \text{ to } 2,577 \text{ miles}$$

6.3 Tests of significance

This topic does not appear on the CIMA syllabus

6.3.1 Basic ideas

Suppose that a battery manufacturer claims that his batteries have a mean lifetime of 600 hours. However, when 100 of his batteries are selected at random, these turn out to have a mean lifetime of only 400 hours. There are two possible explanations for this discrepancy. Either (a) the manufacturer's claim is false or (b) the sample was not representative of the manufacturer's output. The most plausible of these two explanations could be determined by a test of significance. Essentially, this would pose the question: 'If the manufacturer's claim is true, is it likely that a random sample would have given this result?' If the answer is no then the manufacturer's claim must be in doubt.

6.3.2 Terms used in tests of significance

1   The *null hypothesis* (represented by H_0) is the hypothesis to be tested (in this case the claim that the batteries have a mean lifetime of 600 hours). If this hypothesis is 'accepted', then it is assumed that any discrepancy between it and the sample result is owing to sampling error (i.e. the fact that it is unlikely that any sample will perfectly represent the population).
2   The *alternative hypothesis* (H_1) is the hypothesis which is accepted if the null hypothesis is rejected. Acceptance of this hypothesis implies that the difference between the sample result and the H_0 is so great that it is unlikely to be owing to sampling error.
3   *Type I and type II errors*. The decision on whether or not the H_0 should be rejected is based on data from a sample and there is no guarantee that the correct decision will be made. Two possible errors can occur as shown in Table 6.2.

Table 6.2

	Decision	
True situation	Reject H_0	Do not reject H_0
H_0 is true	Type I error	Correct decision
H_0 is false	Correct decision	Type II error

In the examples in this chapter the level of significance has been selected arbitrarily. In practice the choice would depend upon the relative seriousness of type I and type II errors. If a type I error was considered more serious a 1 per cent level might be used. If a type II error was considered to be the most serious a 5 per cent level might be used.

4   The *level of significance* of a test is the probability of rejecting the H_0 when it is true i.e. p(committing a type I error). Conventionally levels of 0.05 (5 per cent) or 0.01 (1 per cent) are employed.
5   *One-tail and two-tail tests*. The hypotheses for the manufacturer could be formulated in two ways.
   In the following formulation, acceptance of the H_1 could imply that either the mean lifetime is above 600 hours or it is below 600 hours. Since there are two possible interpretations this is called a two-tail test.

H_0 mean lifetime, $\mu = 600$ hours
H_1 mean lifetime, $\mu \neq 600$ hours

In contrast, in the formulation below the H_1 is designed specifically to detect whether the mean lifetime is below 600 hours. This is therefore called a one-tail test.

H_0 mean lifetime, $\mu = 600$ hours
H_1 mean lifetime, $\mu < 600$ hours

6.3.3 General procedure for tests of significance

Carrying out a significance test involves the following steps
1   Set up the hypotheses
2   Decide upon the level of significance
3   Formulate the decision rule

4   Perform the calculations
5   Decide whether or not to reject the H_0.

6.3.4 Tests based on the normal distribution

In these tests the normal distribution is used to assess how likely it is that a given sample result will be obtained if the null hypothesis is true. The tests involve the calculation of Z using the following basic formula

$$Z = \frac{\text{sample result} - \text{value specified in } H_0}{\text{appropriate standard error}}$$

To decide whether or not the H_0 should be rejected, Z is compared with a critical value (these are obtained from the normal curve tables). The main critical values and decision rules are shown in Table 6.3. Reject the H_0 if Z has the values indicated.

Table 6.3

	One-tail test		
Level of significance	If H_1 involves $<$ sign	If H_1 involves $>$ sign	Two-tail test
5%	$Z < -1.65$	$Z > 1.65$	$|Z| > 1.96$
1%	$Z < -2.33$	$Z > 2.33$	$|Z| > 2.58$

(a) Testing a hypothesis concerning a population mean

The objective here is to test a hypothesis that the mean of the population has a particular value. To use this test either
1   The sample size must be at least 30, or
2   The population must be normally distributed *and* the population standard deviation must be known.
Z is calculated by using the following formula

$$Z = \frac{\bar{x} - \mu}{\sigma/\sqrt{n}}$$

where $\bar{x}$ = the sample mean, μ = the population mean specified in the H_0, σ = the population standard deviation (if the sample size is at least 30, the sample standard deviation, s, can be used instead) and n = the sample size.

Example

In the week before Christmas it was hoped that the mean takings of a shop's branches would be £40,000. However, forty randomly sampled shops had mean takings of only £37,000 with a standard deviation of £6,000. Have the shops failed to meet the target?

1   Set up hypotheses:

$H_0 \mu = £40,000$
$H_1 \mu < £40,000$ (i.e. a one-tail test)

2   Level of significance: 5 per cent (assume the company are prepared to accept a 5 per cent risk of wrongly rejecting H_0).
3   Decision rule: reject H_0 if Z is less than -1.65.
4   Calculations: $\mu = 40,000, \bar{x} = 37,000, \sigma \approx 6,000, n = 40$,

$$\text{therefore: } Z = \frac{37,000 - 40,000}{\left(\dfrac{6,000}{\sqrt{40}}\right)} = -3.16$$

5   Decision: reject H_0 because -3.16 is less than the critical value. This decision reflects the fact there was less than a 0.05 probability of obtaining this sample result if the target had been reached.

(b) *Testing for a difference between two population means*

This test does not appear on the ICAEW or ICSA syllabuses

This test is designed to ascertain if there is a difference between the mean of one population and the mean of another. A random sample is selected independently from each population. To use this test either,
1   Both samples must be at least size 30, or
2   Both populations must be normally distributed *and* both population standard deviations must be known
Z is calculated as follows

$$Z = \frac{(\bar{x}_1 - \bar{x}_2) - (\mu_1 - \mu_2)}{\sqrt{\dfrac{\sigma_1^2}{n_1} + \dfrac{\sigma_2^2}{n_2}}}$$

where

$\bar{x}_1$ and $\bar{x}_2$ = the means of the samples selected from populations 1 and 2 respectively
$(\mu_1 - \mu_2)$ = the hypothesized difference between the population means. Normally no difference is assumed so that $(\mu_1 - \mu_2) = 0$
σ_1 and σ_2 = the standard deviations of populations 1 and 2 respectively (normally these are approximated by the standard deviation of the samples)
n_1 and n_2 = the sizes of the samples selected from populations 1 and 2 respectively.

Example
Fifty randomly selected marketing managers had a mean annual salary of £20,000 with a standard deviation of £4,000. Seventy randomly selected production managers had a mean annual salary of £19,000 with a standard

deviation of £3,000. Do the mean salaries of the two groups of people differ?

1 Set up hypotheses: $H_0 \, \mu_1 = \mu_2$ (i.e. no difference between means); H_1 $\mu_1 \neq \mu_2$ (a two-tail test because question simply asks: is there a difference?).

2 Level of significance: 5 per cent.

3 Decision rule: Reject H_0 if $|Z|$ exceeds 1.96.

4 Calculations: $\bar{x}_1 = 20{,}000$, $\bar{x}_2 = 19{,}000$, $\sigma_1 \approx 4{,}000$, $\sigma_2 \approx 3{,}000$, $n_1 = 50, n_2 = 70$ therefore:

$$Z = \frac{20{,}000 - 19{,}000}{\sqrt{\dfrac{(4{,}000)^2}{50} + \dfrac{(3{,}000)^2}{70}}} = 1.49$$

5 Decision: Do not reject H_0 because 1.49 does not exceed the critical value. It does not appear that the mean salaries of the two groups differ.

(c) Testing a hypothesis concerning a population proportion

The objective here is to test a hypothesis that a certain proportion of the population has a particular characteristic. Since the test is based on the normal approximation to the binomial distribution it should be used where $n\pi > 5$ (where n is the sample size and π is the population proportion specified in the H_0) and π itself should be between 0.1 and 0.9. Z is calculated as follows

$$Z = \frac{p - \pi}{\sqrt{\dfrac{\pi(1 - \pi)}{n}}}$$

where p is the sample proportion.

Your examination formula sheet may state that the standard error of proportions is

$$\sqrt{\frac{p(1 - p)}{n}}$$

However, π has been substituted here because, in a test of significance, it is the hypothesized proportion and not the sample proportion which is used in the formula.

Example

A company claims that 85 per cent of its orders are delivered within a week. However, in a random sample of 400 orders, only 316 were delivered within this time. Is the company's claim exaggerated?

1 Set up hypotheses: $H_0 \, \pi = 0.85$ (i.e. population proportion $= 0.85$); $H_1 \, \pi < 0.85$ (i.e. a one-tail test).

2 Level of significance: 1 per cent.

3 Decision rule: Reject H_0 if Z is less than -2.33.

4 Calculations: $\pi = 0.85, p = 316/400 = 0.79, n = 400$ therefore:

$$Z = \frac{0.79 - 0.85}{\sqrt{\dfrac{0.85(1 - 0.85)}{400}}} = -3.36$$

5 Decision: Reject H_0 because -3.36 is less than the critical value. It does appear that the claim is exaggerated.

(d) Testing for a difference between two population proportions

This test does not appear on the ICAEW or ICSA syllabuses

This test is designed to ascertain if there is a difference between the proportion of one population having a given characteristic and the proportion of another population having the characteristic. Normally the null hypothesis for this test is that the two population proportions are the same. Therefore, it is first necessary to estimate what this proportion is by taking a weighted average of the two sample proportions (the weights reflecting the sizes of the two samples). Thus:

$$\text{Estimate of population proportion } \bar{p} = \frac{n_1 p_1 + n_2 p_2}{n_1 + n_2}$$

where p_1 = the proportion of items in sample 1 having the characteristic, p_2, the proportion of items in sample 2 and n_1 and n_2 are the two sample sizes. Z is then calculated as follows:

Your examination formula sheet may state that the standard error of the difference in proportions is

$$\sqrt{\frac{p_1(1-p_1)}{n_1} + \frac{p_2(1-p_2)}{n_2}}$$

When (as is usually the case) the null hypothesis is that the two population proportions are the same p_1 and p_2 are both replaced by $\bar{p}$.

$$Z = \frac{p_1 - p_2}{\sqrt{\dfrac{\bar{p}(1-\bar{p})}{n_1} + \dfrac{\bar{p}(1-\bar{p})}{n_2}}} \quad \text{or} \quad \frac{p_1 - p_2}{\sqrt{\bar{p}(1-\bar{p})\left(\dfrac{1}{n_1} + \dfrac{1}{n_2}\right)}}$$

Example

One hundred people in town A are selected at random and thirty own a video recorder. Of seventy randomly selected people in town B, eighteen own a video recorder. Does the proportion of the population owning a video recorder differ between the two towns?

1   Set up hypotheses: $H_0\ \pi_1 = \pi_2$ (i.e. no difference between proportions); $H_1\ \pi_1 \neq \pi_2$ (i.e. a difference exists between proportions).
2   Level of significance: 5 per cent.
3   Decision rule: Reject H_0 if $|Z|$ exceeds 1.96.
4   Calculations: $p_1 = 30/100 = 0.3$, $p_2 = 18/70 = 0.257$, $n_1 = 100$, $n_2 = 70$ therefore:

$$\bar{p} = \frac{100(0.3) + 70(0.257)}{100 + 70} = 0.28$$

$$\text{so } Z = \frac{0.3 - 0.257}{\sqrt{0.28(1 - 0.28)\left(\dfrac{1}{100} + \dfrac{1}{70}\right)}} = 0.61$$

5   Decision: Do not reject H_0 since 0.61 is not greater than the critical value. It does not appear that there is a difference between the two towns in the proportion of people owning video recorders.

6.3.5 Tests based on the t distribution

These tests do not appear on the ICAEW, ICAI or ICSA syllabuses

These tests are used when (a) the samples are small (i.e. less than thirty) and (b) the standard deviation of the population is unknown so that it has to be estimated from the sample and (c) the population is at least approximately normally distributed.

The following basic formula is used to calculate t

$$t = \frac{\text{sample result} - \text{value specified in } H_0}{\text{appropriate standard error}}$$

The decision on whether or not to reject the H_0 is made by comparing t with a critical value which can be obtained from Table 6.1. Note that the table is designed for one-tail tests. For two-tail tests the appropriate value is found by halving the level of significance. For example, for a two-tail test at the 5 per cent level of significance: $(5\%)/2 = 2.5\%$ therefore the $t_{0.025}$ row is used.

The decision rule is: reject the H_0 if t has the value indicated:

	One-tail test	Two-tail test
H_1 involves $<$ sign	H_1 involves $>$ sign	
$t < -$ critical value	$t >$ critical value	$|t| >$ critical value .

(a) The t test for testing a hypothesis concerning a population mean
This test is designed to test a hypothesis that the mean of the population has a particular value. t is calculated as follows

$$t = \frac{\bar{x} - \mu}{\hat{\sigma}/\sqrt{n}}$$

where n = the sample size, $\bar{x}$ = the sample mean, μ = the population mean specified in the H_0, $\hat{\sigma}$ = estimated population standard deviation

$$= \sqrt{\frac{\Sigma(x - \bar{x})^2}{n - 1}}$$

The number of degrees of freedom for this test $= n - 1$.

Example
Boxes of chocolate leaving a production line should weigh 9 ounces. Five boxes selected at random have the following weights. Is there evidence to suggest that the boxes are not meeting the specification?
6 oz, 6 oz, 7 oz, 9 oz, 7 oz

1 Set up hypotheses: H_0 $\mu = 9$ oz (i.e. boxes do meet specification); H_1 $\mu \neq 9$ oz (i.e. a two-tail test).
2 Level of significance: 5 per cent.
3 Decision rule (since this is a two-tail test and the number of degrees of freedom $= 5 - 1 = 4$): Reject the H_0 if $|t|$ exceeds 2.78.
4 Calculations: $\bar{x} = (6 + 6 + 7 + 9 + 7)/5 = 7$

x	6	6	7	9	7
$(x - \bar{x})$	-1	-1	0	2	0
$(x - \bar{x})^2$	1	1	0	4	0

so $\Sigma(x - \bar{x})^2 = 6$
therefore:

$$\hat{\sigma} = \sqrt{\frac{6}{5 - 1}} = 1.22 \text{ oz}$$

$$\text{so } t = \frac{7 - 9}{1.22/\sqrt{5}} = -3.66 \text{ i.e. } |t| = 3.66$$

5 Decision: Reject the H_0 since 3.66 exceeds the critical value. It does appear that the process is not meeting specifications.

(b) The paired t test
This test can be used to ascertain if there is a difference between the means of two populations where the samples from these populations have been matched. For example, an accountant income in one population is compared to an accountant's income in the other population, an architect's income is compared with an architect's and so on. Sometimes the two samples consist of the same people (or objects). For example, the skill of a group of employees could be tested before and after a training course.
In this test

$$t = \frac{\bar{d}}{\hat{\sigma}_d/\sqrt{n}}$$

where:

d = the difference between each matched pair
$\bar{d}$ = the mean difference
$\hat{\sigma}_d$ = the standard deviation of the differences between the pairs
n = the number of *pairs* of observations.

For this test, the number of degrees of freedom $= n - 1$.

Example
The number of miles travelled per gallon of petrol (m.p.g.) by four randomly selected cars is measured under test conditions both before and after a fuel saving device has been fitted. The results are given below. Does the device appear to improve the mean m.p.g. of the cars?

Car	A	B	C	D
m.p.g. before fitting	40	20	33	50
m.p.g. after fitting	50	18	40	55

1 Set up hypotheses: H_0 Mean difference in m.p.g.s $= 0$; H_1 Mean difference in m.p.g.s > 0.
2 Level of significance: 5 per cent.
3 Decision rule (since this is a one-tail test with $4 - 1 = 3$ degrees of freedom): Reject the H_0 if t exceeds 2.35.
4 Calculations: d = m.p.g. after fitting − m.p.g. before fitting

Car	A	B	C	D
d	10	−2	7	5

therefore $\bar{d} = (10 - 2 + 7 + 5)/4 = 5$ m.p.g.
$\hat{\sigma}_d$ is calculated as follows:

d	10	−2	7	5
$(d - \bar{d})$	5	−7	2	0
$(d - \bar{d})^2$	25	49	4	0

so $\Sigma(d - \bar{d})^2 = 78$

which means that $\hat{\sigma}_d = \sqrt{\dfrac{78}{4 - 1}} = 5.09$ m.p.g.

Be careful when calculating this standard deviation: e.g. if $d = -2$ and $\bar{d} = 5$ then $(d - \bar{d}) = -2 - 5 = -7$

thus

$$t = \frac{5}{5.09/\sqrt{4}} = 1.964$$

5 Decision: Do not reject the H_0 since 1.964 does not exceed the critical value. The results do not suggest that the device improves the mean m.p.g.

(c) The t test of the difference between two population means (independent samples)

This test is used to establish whether there is a difference between the means of two populations when (a) small samples are selected independently from each population and (b) the standard deviations of the populations are unknown. To use this test it also needs to be assumed that the two populations are normally distributed with the same standard deviation (even though the value of this standard deviation is unknown).

The value of t is calculated as follows

$$t = \frac{(\bar{x}_1 - \bar{x}_2) - (\mu_1 - \mu_2)}{\hat{\sigma}\sqrt{\dfrac{1}{n_1} + \dfrac{1}{n_2}}}$$

where n_1 and n_2 are the two sample sizes, $\bar{x}_1$ and $\bar{x}_2$ are the two sample means, $(\mu_1 - \mu_2)$ is the hypothesized difference between the two sample means (normally this is 0) and $\hat{\sigma}$ is the estimate of the standard deviation of the populations. Since this is assumed to be the same for the two populations, a pooled estimate of its value, based on the two sample standard deviations (s_1 and s_2) is calculated as follows

$$\hat{\sigma} = \sqrt{\frac{(n-1)s_1^2 + (n-1)s_2^2}{n_1 + n_2 - 2}}$$

The number of degrees of freedom for this test is $n_1 + n_2 - 2$.

Example

A company operates two types of road vehicle (A and B). A random sample of three type A vehicles had mean operating costs in a given month of £220 with a standard deviation of £75.50. A random sample of four type B vehicles had mean operating costs during the month of £340 with a standard deviation of £145.10. Do these results suggest a difference in the mean costs of operating the vehicles?

1 Set up hypotheses: $H_0\ \mu_1 = \mu_2$ (i.e. the mean costs do not differ); H_1 $\mu_1 \neq \mu_2$ (i.e. the mean costs differ).
2 Level of significance: 5 per cent.
3 Decision rule (since there are $3 + 4 - 2 = 5$ degrees of freedom and this is a two-tail test): Reject H_0 if $|t|$ exceeds 2.57.
4 Calculations: $\bar{x}_1 = £220$, $\bar{x}_2 = £340$, $s_1 = £75.50$, $s_2 = £145.10$ so that

$$\hat{\sigma} = \sqrt{\frac{(3-1)(75.5)^2 + (4-1)(145.1)^2}{3+4-2}} = £122.10$$

$$\text{thus } t = \frac{(220 - 340) - 0}{122.1\sqrt{\frac{1}{3} + \frac{1}{4}}} = -1.29 \text{ so } |t| = 1.29$$

5 Decision: Do not reject H_0 since 1.29 does not exceed the critical value. There does not appear to be a difference in the mean operating costs of the vehicles.

6.3.6 Tests based on the chi-squared distribution

These tests do not appear on the ICAEW or ICSA syllabuses

The chi-squared (χ^2) tests enable one to compare the observed frequency of occurrence of events with the frequencies which would be expected to occur if the null hypothesis is true. For example, if the null hypothesis is that a coin is fair, twenty-five heads would be expected in fifty tosses of the coin. However, suppose that, when the coin is tossed, twenty-nine heads are observed. Is this difference between the observed and expected frequencies sufficiently large for the null hypothesis to be rejected? The χ^2

test provides a systematic procedure for answering this type of question. The test involves the calculation of χ^2 where:

$$\chi^2 = \Sigma \frac{(O - E)^2}{E}$$

where O = observed frequencies, E = expected frequencies. If the χ^2 value exceeds the appropriate critical value then the null hypothesis is rejected. The critical value can be obtained from Table 6.4 and depends upon the number of degrees of freedom (these are dealt with later).

Table 6.4 Chi-squared distribution

Degrees of freedom	1	2	3	4	5
5% level of significance	3.8	6.0	7.8	9.5	11.1
1% level of significance	6.6	9.2	11.3	13.3	15.1

Note that if any of the expected values are less than five, classes should be merged (see the second example below).

(a) Chi-squared tests on contingency tables

Consider the following table (known as a contingency table) which summarizes the results of a survey of 200 of a company's customers.

Type of customer	Speed of payment of last invoice (No. of customers)			
	under 2 months	2 to 6 months	over 6 months	Total
Wholesaler	28	32	22	82
Retailer	84	16	18	118
Total	112	48	40	200

The company want to find if there is an association between the speed of payment and the type of customer. The test is conducted as follows.

1 Formulate hypotheses: H_0 There is no association between speed of payment and customer type; H_1 There is an association between speed of payment and customer type.
2 Level of significance: 5 per cent.
3 Decision rule: For a contingency table, no. of degrees of freedom = (no. of rows − 1) (no. of columns − 1). In this case there are $(2 − 1)(3 − 1) = 2$ degrees of freedom. Therefore at the 5 per cent level of significance the critical value = 6.0. The H_0 should be rejected if χ^2 exceeds this value.
4 Calculations: First the expected frequencies need to be calculated for each cell in the table (i.e. the frequencies which would be expected if

Do not count the 'total' row and column when calculating the number of degrees of freedom. An even easier way of calculating the number of degrees of freedom is simply to cross out the bottom row and the last column (again ignoring the 'total' row and column) and count the number of cells which are left.

the null hypothesis is true). This can be done using the following formula

$$E = \frac{\text{row total} \times \text{column total}}{\text{grand total}}$$

For example, to calculate the expected number of wholesalers who paid within two months: the total of the wholesaler row = 82, the total of the 'under 2 months' column is 112 and the grand total for the table is 200. Therefore

$$E = \frac{82 \times 112}{200} = 45.92$$

The Es can take on fractional values because they represent the mean number of customers which we would expect to occur in the various parts of the table if the sampling process were repeated a large number of times.

The expected frequencies are shown below, in parentheses, next to the observed frequencies:

Type of customer	Speed of payment of last invoice (No. of customers)			
	under 2 months	2 to 6 months	over 6 months	Total
Wholesaler	28 (45.92)	32 (19.68)	22 (16.40)	82
Retailer	84 (66.08)	16 (28.32)	18 (23.60)	118
Total	112	48	40	200

Therefore $\chi^2 = \dfrac{(28 - 45.92)^2}{45.92} + \dfrac{(32 - 19.68)^2}{19.68} + \dfrac{(22 - 16.40)^2}{16.40}$

$$+ \frac{(84 - 66.08)^2}{66.08} + \frac{(16 - 28.32)^2}{28.32} + \frac{(18 - 23.60)^2}{23.60} = 28.16$$

5   Decision: Reject H_0 since 28.16 exceeds the critical value. It does appear that speed of payment is associated with customer type. Inspection of the table suggests that wholesalers tend to be slower payers.

Note that the chi-squared test should only be carried out on frequencies and never on percentages. For example, if the contingency table had shown that 14 per cent of the sample were wholesalers who paid in under two months, it would have been necessary to convert this to a frequency (i.e. 14 per cent of 200 = 28).

(b) Goodness of fit tests

Chi-squared tests can be used to test whether a particular probability distribution fits a set of data. In these tests the null hypothesis is always that the probability distribution *does* fit the data.

The number of degrees of freedom is given by

No. of classes for which E values are calculated (after any necessary merging)
− no. of parameters which have to be estimated from the data − 1

Strictly speaking, when there is only one degree of freedom in a chi-squared test, the following formula should be used:

$$x^2 = \Sigma \frac{(|0 - E| - \tfrac{1}{2})^2}{E}$$

Example

The frequency distribution below shows the number of machine breakdowns occurring during shifts at a factory. Does the data suggest that the breakdowns follow a Poisson distribution?

No. of breakdowns	(x)	0	1	2	3	4 or more
No. of shifts	(f)	20	30	10	8	2

(i.e. twenty shifts had no breakdowns, thirty had one breakdown etc.)

1   Formulate hypotheses: H_0 The breakdowns do follow Poisson distribution; H_1 The breakdowns do not follow a Poisson distribution.
2   Level of significance: 5 per cent.
3   Calculations: Mean number of breakdowns is

$$\frac{\Sigma fx}{\Sigma f} = \frac{(0 \times 20) + (1 \times 30) + (2 \times 10) + (3 \times 8) + (4 \times 2)}{70} = 1.2 \text{ breakdowns}$$

The probabilities of 0, 1, 2, 3 and 4 or more breakdowns, assuming a Poisson distribution with a mean of 1.2, are next calculated using the formula or tables. The probabilities are:

No. of breakdowns	0	1	2	3	4 or more
Probability	0.301	0.361	0.217	0.090	0.031

This means that, for example, 0 breakdowns would be expected in 30.1 per cent of shifts. The expected frequencies are thus obtained by multiplying the probabilities by the total frequency (i.e. 70) to give

No. of breakdowns	0	1	2	3	4 or more
Expected frequency	21.1	25.3	15.2	6.3	2.2

However, since the expected frequency in the last class is less than five the last two classes must be merged to form a '3 or more class' with expected frequency $6.3 + 2.2 = 8.5$. The observed and expected frequencies are now

No. of breakdowns	0	1	2	3 or more
Observed frequencies	20	30	10	10
Expected frequencies	21.1	25.3	15.2	8.5

For this data the χ^2 formula gives a value of 3.0 (approximately).

4   Decision rule: One parameter (the mean) had to be estimated from the data and there were four E values after merging the classes. Therefore the number of degrees of freedom $= 4 - 1 - 1 = 2$ so the critical value is 6.0.
5   Decision: Do not reject the H_0. The Poisson distribution does appear to be a suitable model for the machine breakdowns.

Note that, in testing for a normal distribution, two parameters (the mean and standard deviation) would have to be estimated from the data. In testing for a binomial distribution, one parameter (p, the probability of an event occurring in a single trial) would have to be estimated.

7

Functions and graphs

The material in this chapter is relevant to the following syllabuses: CIMA, ICAEW, CACA(1.5), ICSA

7.1 Introduction

This chapter considers a number of techniques which are used in mathematical modelling. Mathematical models can provide concise and unambiguous descriptions of decision problems. Moreover, they enable problems to be explored and investigated through mathematical analysis. However, a model will almost certainly be a simplification of the real problem and therefore any 'solution' suggested by the model should be viewed with caution.

7.2 Functions

In many situations the value of one variable (the dependent variable) depends upon the value of another variable (the independent variable). For example, the sales of a product may depend upon its price. When variables are related in this way, the dependent variable is said to be a function of the independent variable. Thus if S represents the sales of a product and p its price then S is a function of p which can be written as: $S = f(p)$.

7.2.1 Linear functions

Suppose that a product is manufactured in batches. Each time a batch is manufactured it costs £50 to set up the machinery for the production run and an additional £10 for each unit produced. If y is the total cost of a production run (in pounds) and x the number of units produced, the relationship can be written as

$$y = 50 + 10x$$

The table below shows the total cost of production runs of various sizes.

No. of units produced	0	1	2	3	4
Total cost (£)	50	60	70	80	90

When these points are plotted on a graph (Figure 7.1) they all fall on a straight line. This is, therefore, an example of a linear function or relationship. Note also that the line passes at a height of £50 above the origin. This distance, which represents fixed costs, is known as the *intercept*. It can also be seen that the line has a constant slope (i.e. its steepness never changes). The slope of a line can be measured by the following ratio

Note that conventionally, the vertical axis represents the dependent variable.

$$\frac{\text{Vertical distance between any two points}}{\text{Horizontal distance between these points}}$$

If the line slopes downwards from left to right this slope is negative.
Thus using the points labelled *l* and *m* on the graph

$$\text{Slope} = \frac{90 - 50}{4 - 0} = 10$$

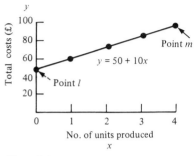

Figure 7.1

This reflects the fact that every extra unit produced increases total costs by £10. In this case, therefore, the slope represents variable costs.

All linear relationships can be represented by equations of the form $y = a + bx$ (where a and b are numbers). Comparing this with $y = 50 + 10x$, it can be seen that a is the intercept and b (the number multiplied by x) is the slope.

Example

Find the intercept and slope of the line $y = 8 - 3x$ and plot the line on a graph.

Here, a = the intercept = 8 and b = the slope = −3. To plot the line only two points through which it passes are needed. When $x = 0$, $y = 8$ (i.e. the intercept) and when $x = 4$, $y = 8 - 3(4) = -4$. The line is plotted in Figure 7.2. Note that the negative slope implies that as x increases, y decreases.

7.2.2 *Quadratic functions*

Some relationships appear as curves when plotted on a graph and these are described as curvilinear relationships. A particular type of curvilinear relationship is known as a quadratic function. The equations of quadratic functions have the general form $y = ax^2 + bx + c$ where a, b and c are numbers. Thus, for the equation $y = 2x^2 + 3x + 4$, a is 2, b is 3 and c is 4 while for $y = x^2$, a is 1 and b and c are zero.

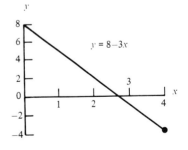

Figure 7.2

Example

In the manufacture of a product the daily profit earned (in pounds) (y) is related to the number of units produced (x) in the following way: $y = -10x^2 + 300x - 1,250$. Plot the graph of the equation.

To plot the graph several points which the curve passes through need to be calculated.

For example at $x = 0$, $y = -10(0)^2 + 300(0) - 1.250 = -1,250$; and at $x = 5$, $y = -10(5)^2 + 300(5) - 1,250 = 0$ etc.

The following table gives the value of y for values of x between zero and thirty.

Output (no. of units)	x	0	5	10	15	20	25	30
Profit (£)	y	−1,250	0	750	1,000	750	0	−1,250

The graph is shown in Figure 7.3. It can be seen that the maximum profit is made at an output of fifteen units. All quadratic functions have either maximum or minimum points. Note that the curve is symmetrical about the maximum point. No profit is made at production levels of five and twenty-five units so these are the break-even levels of output. These two points, where the line crosses the x axis, are also known as the *roots of the quadratic*. Since y equals 0 at these points they could be calculated without drawing a graph by solving the equation $-10x^2 + 300x - 1,250 = 0$.

The following formula can be used to solve equations of this type

The nature of the curve can be ascertained quickly as follows. If a is positive the quadratic will have a minimum point. If it is negative the function will have a maximum point.

± means 'plus or minus' i.e. 'first add and then subtract the numbers' to get the two results.

$$x = \frac{-b \pm \sqrt{b^2 - 4ac}}{2a}$$

In this case $a = -10$, $b = 300$ and $c = -1,250$. Thus

Alternatively the equation could have been factorized so that $(-5x + 25)(2x - 50) = 0$ giving $x = 5$ or 25.

$$x = \frac{-300 \pm \sqrt{(300)^2 - 4(-10)(-1,250)}}{2(-10)} = \text{either 5 or 25}$$

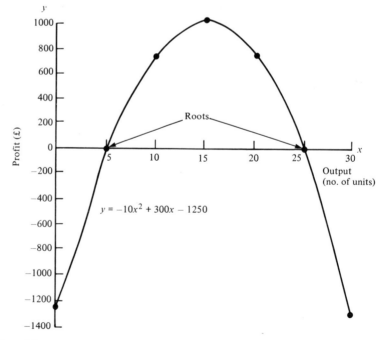

$y = -10x^2 + 300x - 1250$

Figure 7.3

Not all quadratic equations have roots: many 'turn back' before crossing the x axis. When this is the case the above formula would involve the square root of a negative number for which there is no real solution.

7.3 Differential calculus

In some problems it is necessary to measure the slope of a curve. For example if a curve represents the relationship between total costs and output, its slope will represent the rate at which total costs are increasing (i.e. marginal costs). However, while a straight line has a constant slope, the slope of a curve is constantly changing and the slope must therefore be measured at a specific point. This measurement can be made by using a process called differentiation. For a function $y = f(x)$ the slope is represented by the symbol dy/dx. This is also called the *derivative*, or *differential coefficient*, of y with respect to x.

Note that dy/dx does not mean d multiplied by y or x or even dy divided by dx. The expression is simply a symbol to represent the slope of a function. You may come across other symbols which mean the same as dy/dx such as

$$\frac{d(y)}{dx}, \quad D_x(y), f'(x) \text{ and } y'.$$

7.3.1 Differentiating simple functions

(a) Differentiating a constant
The differential coefficient of a constant is zero. For example if $y = 50$ $dy/dx = 0$. This is because y never changes. Plotted on a graph (Figure 7.4(a)) the function is a horizontal line so its slope is zero.

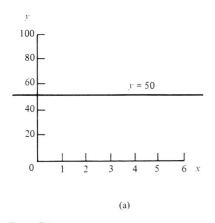

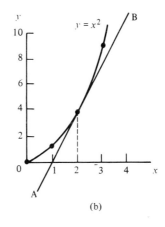

(a)   (b)

Figure 7.4

(b) Differentiating $y = x^n$
If $y = x^n$ (where n is a number) then it can be shown that $dy/dx = nx^{n-1}$. The slope at a specific point is found by substituting the appropriate value of x.

Example
Differentiate the following equations and find their slope at $x = 2$
(i) $y = x^2$ (ii) $y = x^3$ (iii) $y = x^{-2}$

x^{-2} *means* $\dfrac{1}{x^2}$

(i) If $y = x^2$, $dy/dx = 2x^{2-1} = 2x$. At $x = 2$, the slope $= dy/dx = 2(2) = 4$. This is shown in Figure 7.4(b) where the tangent AB represents the slope of the curve at $x = 2$. This tangent has a slope of 4.

(ii) If $y = x^3$, $dy/dx = 3x^{3-1} = 3x^2$.
At $x = 2$, the slope $= dy/dx = 3(2)^2 = 12$.

(iii) If $y = x^{-2}$, $dy/dx = -2x^{-2-1} = -2x^{-3}$.
At $x = 2$, the slope $= dy/dx = -2(2)^{-3} = -2/8 = -0.25$.

(c) Differentiating ax^n

If $y = ax^n$ (where a is a number) then $dy/dx = nax^{n-1}$

Example

Differentiate the following equations and find their slope at $x = 2$.
(i) $y = 2x^3$ (ii) $y = -2x^{1/2}$.

$x^{1/2}$ means $\sqrt{x}$ while $x^{-1/2}$ means $\dfrac{1}{\sqrt{x}}$

(i) If $y = 2x^3$, $dy/dx = 3(2)x^{3-2} = 6x$.
At $x = 2$, the slope $= dy/dx = 6(2) = 12$.

(ii) If $y = -2x^{1/2}$, $dy/dx = (1/2)(-2)x^{1/2-1} = -x^{-1/2}$.
At $x = 2$ the slope $= -(2)^{-1/2} = -1/(\sqrt{2}) = -0.707$.

7.3.2 Differentiating a sum

When an equation involves a number of terms added together (or subtracted from one another) the differentiation is carried out term by term. Thus if $y = x^4 + 2x^3 - 2x^2$, $dy/dx = 4x^3 + 6x^2 - 4x$.

Thus at $x = 2$ the curve would have a slope of $4(2)^3 + 6(2)^2 - 4(2) = 48$.

Example

Differentiate $y = 50 + 10x$.

Remember that a number raised to the power 0 equals 1. Also $x^1 = x$.

$y = 50 + 10x$. Therefore $dy/dx = 0 + 10x^0 = 10(1) = 10$. The equation which has been differentiated is the linear relationship used in Section 7.2.1. This confirms that the slope of the line is constant and is equal to 10.

7.3.3 Finding maximum and minimum points

Differential calculus can be used to find the maximum or minimum points (also known as turning points) of a curve (e.g. the level of output where costs are minimized or profit maximized). This is because the slope of a curve at a turning point is zero. Finding the turning points involves the following steps:

1 Differentiate the function.
2 Set the derivative to zero and solve for x to establish where the turning point is.
3 Differentiate the function obtained in 1 (this gives the second differential coefficient: d^2y/dx^2). If the second differential coefficient is positive at the turning point then the point is a minimum. If it is negative then the point is a maximum.

If the second differential coefficient is zero then a point of inflexion exists. This is where a 'twist' occurs in the curve before it continues to increase or decrease.

Example

The following equation represents the relationship between the average cost in pounds of producing a product (C) and the output of the product (x). Identify any minimum or maximum points.

$$C = 3x^2 - 60x + 500$$

1 Differentiating the function: $dC/dx = 6x - 60$.
2 At the turning point the slope is 0 thus $6x - 60 = 0$, therefore a turning point exists at $x = 10$.
3 To determine whether the point is a minimum or maximum point it is necessary to differentiate $6x - 60$ to give the second differential coefficient. Second differential coefficient is

$$\frac{d^2C}{dx^2} = 6$$

Since the second differential coefficient is positive average costs are minimized at an output level of ten units. These minimum average costs can be found by substituting $x = 10$ into the original equation. Thus at $x = 10$: average costs $= C = 3(10)^2 - 60(10) + 500 = £200$.

If d^2C/dx^2 had been a function of x, for example if $d^2C/dx^2 = 6x - 8$, then $x = 10$ would have been substituted into the expression to establish whether d^2C/dx^2 was positive or negative. Thus d^2C/dx^2 would have equalled $6(10) - 8 = 52$ which, since it is positive, would again have indicated a minimum point.

7.4 Integration

This topic is not on the ICSA syllabus

Integration is the reverse of differentiation. For example, suppose that for a curve it is known that $dy/dx = 6x^2$, but the original equation of the curve has been lost. To find the original equation it is necessary to integrate $6x^2$ with respect to x. This is written as:

$$\int 6x^2 \, dx$$

Now it may seem that

$$\int ax^n \, dx = \frac{ax^{n+1}}{n+1}$$

therefore

$$\int 6x^2 \, dx = \frac{6x^{2+1}}{2+1} = \frac{6x^3}{3} = 2x^3$$

Therefore it appears that the original equation was $y = 2x^3$. However, all of the following equations could also have led to $dy/dx = 6x^2$: $y = 2x^3 + 3$, $y = 2x^3 + 10$, $y = 2x^3 - 45$ etc. This is because the constant disappears in differentiation and, without further information, it is impossible to determine what its value was. Therefore $\int 6x^2 \, dx = 2x^3 + C$ where $C =$ the unknown constant. Because C is unknown this is called an indefinite integral. Thus:

$$\int ax^n \, dx = \frac{ax^{n+1}}{n+1} + C$$

Note, however, that this rule for integration will not work when $n = -1$, where a different method has to be used.

Example

The equation for the slope of a total cost curve (i.e. the marginal cost) is given below where T = total cost and x = output.

$$dT/dx = 4x - 1$$

Find the equation of the original total cost curve.

$$T = \int (4x - 1)\, dx = \int (4x^1 - 1x^0)\, dx \text{ (since } x^1 = x \text{ and } x^0 = 1)$$

$$= \frac{4x^{1+1}}{1+1} - \frac{1x^{0+1}}{0+1} + C$$

You may find it simpler to remember that the integral of a constant with respect to x is the constant multiplied by x. For example

$$\int 1 dx = x + C, \text{ and } \int 10 dx = 10x + C$$

It is always a good idea to differentiate the expression you have obtained to check that it results in the original expression: e.g. if $T = 2x^2 - x + C$ then $dT/dx = 4x - 1$, which is correct.

so that the original equation is: $T = 2x^2 - x + C$. If it is known that at a zero level of output, total costs = £100 (i.e. fixed costs are £100) then $C = 100$ so that $T = 2x^2 - x + 100$ would be the equation.

7.5 Finding points of intersection

It is often necessary to find the point of intersection of lines or curves in order to determine, for example, the point where demand equals supply or where total revenue equals total cost (i.e. the break-even point). The point of intersection can be found by solving the equations simultaneously.

Example

Find the point of intersection of the lines: $2x + 3y = 10$ (equation 1) and $4x + y = 5$ (equation 2).

To solve the equations, one of the two variables must be eliminated. One way of doing this is to multiply equation 1 by 2. This ensures that, when the second equation is subtracted from the first, x will disappear.

2 × equation 1 is: $4x + 6y = 20$
equation 2 is, as before, $4x + y = 5$
_____

2 × equation 1 − equation 2 is $0 + 5y = 15$ so that $y = 3$

Substituting $y = 3$ into equation 2 gives $4x + 3 = 5$ so that $4x = 2$ i.e. $x = 0.5$. Therefore the lines intersect at the point $x = 0.5$, $y = 3$.

8
Matrix algebra and financial mathematics

The material in this chapter is relevant to the following syllabuses, except where otherwise indicated: CIMA, ICAEW, ICAS, CACA(1.5), ICSA

8.1 Introduction

The formulation and solution of many problems are made easier by using matrices and the first half of this chapter defines and examines the basic operations of matrix algebra. The second part of the chapter deals with the applications of mathematics to financial problems such as investment appraisal.

8.2 Matrices

Matrices do not appear on the ICAEW, ICAS or ICSA syllabuses

A matrix is an array of numbers (which are called the elements of the matrix). For example, the table below shows the probabilities of readers switching between two local newspapers from one day to the next (i.e. if a reader bought newspaper A today, there is a 0.7 probability that he will also buy A tomorrow, a 0.1 probability that he will switch to B tomorrow, etc.).

		Paper purchased tomorrow		
		A	B	None
Paper purchased	A	0.7	0.1	0.2
today	B	0.3	0.6	0.1
	None	0.2	0.4	0.4

This table can be represented by the matrix Y

$$Y = \begin{pmatrix} 0.7 & 0.1 & 0.2 \\ 0.3 & 0.6 & 0.1 \\ 0.2 & 0.4 & 0.4 \end{pmatrix}$$

The order of a matrix describes its size and is written as: no. of rows × no. of columns. Thus Y is a 3×3 matrix. It is also called a square matrix because it has the same number of rows and columns. Matrix X, below, is of order 2×4

$$X = \begin{pmatrix} 4 & 1 & 6 & 8 \\ 0 & 3 & 7 & 2 \end{pmatrix}$$

When a matrix has only one row or one column, e.g. Z = (2 3 0 6), it is called a *vector*.

8.2.1 Addition and subtraction of matrices

It is only possible to add or subtract matrices which are of the same order. Addition and subtraction simply involve adding corresponding elements together or subtracting corresponding elements from each other as shown in the examples below

$$\begin{pmatrix} 2 & 6 \\ 3 & 7 \\ 4 & 8 \end{pmatrix} + \begin{pmatrix} 3 & 7 \\ 5 & 2 \\ 9 & 1 \end{pmatrix} = \begin{pmatrix} 2+3 & 6+7 \\ 3+5 & 7+2 \\ 4+9 & 8+1 \end{pmatrix} = \begin{pmatrix} 5 & 13 \\ 8 & 9 \\ 13 & 9 \end{pmatrix}$$

$$\begin{pmatrix} 7 & 4 \\ 3 & 2 \end{pmatrix} - \begin{pmatrix} 1 & 3 \\ 0 & 5 \end{pmatrix} = \begin{pmatrix} 7-1 & 4-3 \\ 3-0 & 2-5 \end{pmatrix} = \begin{pmatrix} 6 & 1 \\ 3 & -3 \end{pmatrix}$$

8.2.2 Multiplication of a matrix by a scalar

An ordinary number is called a scalar in matrix algebra. Multiplying a matrix by a scalar simply involves multiplying every element of the matrix by the scalar. For example

$$2 \times \begin{pmatrix} 3 & -1 \\ 6 & 7 \end{pmatrix} = \begin{pmatrix} 2 \times 3 & 2 \times -1 \\ 2 \times 6 & 2 \times 7 \end{pmatrix} = \begin{pmatrix} 6 & -2 \\ 12 & 14 \end{pmatrix}$$

8.2.3 Multiplication of two matrices

When multiplying two matrices together the order of multiplication is important. For example, P × Q will generally give a different result to Q × P. Also, the multiplication can only be carried out if the number of columns in the first matrix is the same as the number of rows in the second. Thus a 4 × 7 matrix can be multiplied by a 7 × 3 matrix. This would result in a 4 × 3 matrix as shown below

Order of resulting matrix

$$(4 \times \textcircled{7}) \qquad (\textcircled{7} \times 3)$$

These are equal so
multiplication is possible

The process of multiplication is demonstrated below:

Example
Find A × B where

$$A = \begin{pmatrix} 3 & 2 & 4 \\ 6 & 8 & 1 \end{pmatrix} \qquad \text{and B} = \begin{pmatrix} 9 & 7 \\ 5 & 0 \\ 12 & 10 \end{pmatrix}$$

Here a 2×3 matrix is to be multiplied by a 3×2 matrix so a 2×2 matrix will result. Therefore four elements need to be calculated. The element in row 1, column 1 of the resulting matrix is obtained by multiplying the elements of row 1 of A by the elements of column 1 of B, as shown below, and adding up the resulting figures

Note that, in each case, the multiplication is set out as follows: 1st element in A's row × 1st element in B's column
2nd element in A's row × 2nd element in B's column etc.

$$3 \times \ 9 = 27$$
$$2 \times \ 5 = 10$$
$$4 \times 12 = 48$$
$$\overline{}$$
$$85$$
$$\overline{}$$

Matrix so far: $\begin{pmatrix} 85 & ? \\ ? & ? \end{pmatrix}$

To obtain the element in row 1, column 2 of the resulting matrix, the elements in row 1 of A are multiplied by the elements in column 2 of B, as shown below, and the resulting figures are summed

$$3 \times \ 7 = 21$$
$$2 \times \ 0 = \ 0$$
$$4 \times 10 = 40$$
$$\overline{}$$
$$61$$

Matrix so far: $\begin{pmatrix} 85 & 61 \\ ? & ? \end{pmatrix}$

The element in row 2, column 1 of the resulting matrix is obtained in a similar way by multiplying row 2 of A by column 1 of B which gives a value of 106. Also, multiplying row 2 of A by column 2 of B gives the element in row 2 column 2 in the resulting matrix. Its value is 52. Therefore, the final matrix is

$$\begin{pmatrix} 85 & 61 \\ 106 & 52 \end{pmatrix}$$

Thus the rule is: to obtain the element in row x column y of the resulting matrix, multiply the elements in row x of the first matrix by the elements in column y of the second matrix (as shown above) and then sum the results.

8.3 Financial mathematics

The rest of this chapter looks at the application of mathematics in financial calculations such as compound interest, discounting and investment appraisal.

8.3.1 Arithmetic progressions

This topic does not appear on the ICAEW, ICAS, CACA(1.5) or ICSA syllabuses

The numbers 2, 6, 10, 14, 18 are said to be in an arithmetic progression because each term is equal to the previous term plus 4. The difference

The numbers in an arithmetic progression can get smaller, e.g. 8, 2, −4, −10 have a common difference of −6.

between the successive terms (i.e. 4 in this case) is called the common difference.

The nth term in an arithmetic progression $= a + (n-1)d$ where a is the first term and d is the common difference.

The sum of the first n terms of an arithmetic progression, S_n, can be calculated using the formula

$$S_n = \frac{n}{2}[2a + (n-1)d]$$

Example

In a contract the annual rent paid for an item will increase by £6 per year. The rent paid in the first year is £20. Find the rent payable in the fifth year and the total rent to be paid over the first five years.

$a = 20$, $n = 5$ and $d = 6$ so the fifth year rent $= 20 + (5-1)6 = £44$. The total rent paid for the first five years is

$$\frac{5}{2}[2(20) + (5-1)6] = £160$$

8.3.2 Geometrical progressions

This topic does not appear on the ICAEW, ICAS, CACA(1.5) or ICSA syllabuses

The numbers 2, 6, 18, 54, 162 are said to be in a geometrical progression because each term is equal to the preceding term multiplied by 3. The ratio of each term to its predecessor (i.e. 3 in this case) is called the common ratio.

R has been used here to avoid confusion with the rate of interest, which is generally denoted by r.

The nth term in a geometrical progression is aR^{n-1} where a is the first term and R is the common ratio. The sum of the first n terms of a geometrical progression, S_n, can be calculated by using the formula

$$S_n = \frac{a(1-R^n)}{(1-R)}$$

Example

The annual cost of maintaining a machine is £200 in its first year but this will increase by 20 per cent per year. Find the cost of maintaining the machine in year 4 and also the total cost of maintaining the machine over four years.

R = 1.2 here because if numbers increase by 20 per cent each number is 1.2 times its predecessor.

$a = £200$, $R = 1.2$ and $n = 4$, so the fourth year cost will be $200(1.2)^{4-1} = £345.60$. The total maintenance cost over the four years is

$$\frac{200(1-(1.2)^4)}{1-1.2} = £1,073.60$$

If the common ratio is a fraction (e.g. 10, 5, 2.5, 1.25 where $R = 0.5$) then it is possible to find the sum of an infinite number of terms (the sum to infinity), S where

$$S = a/(1 - R)$$

Thus, for this example, $S = 10/(1 - 0.5) = 20$.

8.3.3 Compound interest

Bank and building society accounts generally earn compound interest. At the end of each period, interest is added to the account and in the next period interest is therefore earned on the original amount invested plus the interest added in previous periods. The sum of money in an account after n periods, S, can be found by using the following formula:

$$S = P(1 + r)^n$$

where P is the principal i.e. the original amount invested and r is the rate of interest per period expressed as a proportion (i.e. if 5 per cent is the rate of interest $r = 0.05$).

Example
£500 is invested today in an account which earns compound interest at 10 per cent per annum. How much money will be in the account after six years (assuming that no withdrawals or additional deposits are made)?

Most scientific calculators have an x^y button which can be used to perform calculations like $(1.1)^6$.

$P = £500, r = 0.1$ and $n = 6$ therefore $S = 500(1 + 0.1)^6 = £885.78$.

8.3.4 Discounting and present values

Discounting can be used to answer the question: 'To receive a certain sum of money in n years' time, how much needs to be invested now?' This is the reverse of the problem considered in the previous section since S, the sum of money to be received in the future, is known, but P the amount to be invested needs to be calculated. The compound interest formula can be manipulated to give

$$P = \frac{S}{(1 + r)^n}$$

P is known as the present value of the sum of money to be received in the future.

Example
How much money needs to be invested now at a rate of interest of 10 per cent to receive £200 in three years' time?

$S = £200, r = 0.1$ and $n = 3$ so

$$P = \frac{200}{(1 + 0.1)^3} = £150.26$$

Thus the present value of £200 to be received in three years' time is £150.26.

Alternatively, present value tables (e.g. Table 11 in the CIMA's *Mathematical tables for Students*) can be used here. In this case, $P = S \times$ the present value (PV) factor obtained from tables. At a rate of interest of 10 per cent and with n equal to 3 the CIMA tables give a PV factor of 0.75. Therefore $P = 200 \times 0.75 = £150$ (the two answers differ because the figures in the tables have been rounded).

8.3.5 Net present value

£100 due to be received in a year's time is worth less than £100 held today. This is because the £100 held today can be invested and will have earned interest by the end of the year. Therefore, when evaluating the future returns from an investment it is necessary to take into account the period which will elapse before the money will be received. This can be done by discounting all future cash flows to their present value (the result is called the discounted cash flow or DCF). In the net present value (NPV) approach to investment decisions the net cash flows (i.e. inflows – outflows) occurring in each period are discounted and then summed.

Example

A company is thinking of investing £20,000 in the purchase of a machine which will have a lifetime of three years. The outflows and inflows of cash resulting from the purchase of the machine are given below for each of the three years. (It is assumed that these cash flows occur at the end of each year.) If the company could earn 10 per cent interest by investing the £20,000 elsewhere, should they purchase the machine?

Most exam. questions assume that the cash flows occur at the year end. Otherwise the calculations are rather complex!

Year	0 (i.e. now)	1	2	3
Cash inflows	£0	£10,000	£10,000	£5,000
Cash outflows	−£20,000	£0	£0	£0

Because the company could earn 10 per cent interest elsewhere this becomes the discount rate and r is set equal to 0.1. The calculations can be set out as shown in Table 8.1 (note that tables could have been used in column 5).

The positive net present value implies that the £20,000 invested in the machine will have a return of more than 10 per cent (i.e. more than could be gained by investing elsewhere). This suggests that the machine should be purchased. A negative NPV would have indicated that money invested in the machine would earn less than the 10 per cent which could be gained elsewhere. Note, however, that the cash flows and the length of life of the machine are only estimates and no attempt has been made to take into account the uncertainty associated with the cash flows by, for example, using probabilities.

Table 8.1

End of year	Cash inflow	Cash outflow	Net cash flow	Discounted cash flow	
0	£0	−£20,000	−£20,000		−£20,000
1	£10,000	£0	£10,000	$\dfrac{10,000}{(1+0.1)^1} =$	£9,091
2	£10,000	£0	£10,000	$\dfrac{10,000}{(1+0.1)^2} =$	£8,264
3	£5,000	£0	£5,000	$\dfrac{5,000}{(1+0.1)^3} =$	£3,757
			Net present value		£1,112

8.3.6 The internal rate of return

Another way of evaluating a proposed investment is to use the internal rate of return (IRR) method. The IRR is the discount rate which leads to a net present value of zero and it can be calculated approximately by using a graphical approach. For the example discussed in the previous section it was shown that a discount rate of 10 per cent leads to an NPV of £1,112. If the discount rate had been 12 per cent this would have led to an NPV of £460. A 14 per cent discount rate leads to an NPV of −£158. This suggests that the IRR lies between 12 per cent and 14 per cent. Plotting these points on a graph (Figure 8.1) and interpolating suggests that the IRR is about 13.5 per cent.

An investment is worth while if its IRR exceeds an acceptable or target figure. Thus if the company in the above example wish to earn at least a 10 per cent return on their money then it is worth investing in the machine since it has an IRR above this figure.

In some cases the IRR may not exist while in other (rare) cases it may have more than one value. Also, projects can have an IRR which is a negative percentage.

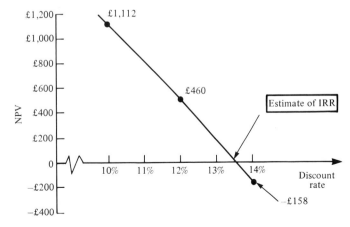

Figure 8.1

8.3.7 The present value of an annuity

An annuity is a fixed sum of money which is usually paid annually over a number of years. The present value of an annuity can be calculated by using special tables (e.g. Table 12 Cumulative Present Value of £1 in the CIMA tables). Alternatively, all the cash flows could be discounted individually and summed, but this is slow and tedious where a large number of years is involved.

Example

An individual is considering selling a piece of land in return for a payment of £200 which will be received at the end of each year for the next ten years. Calculate the present value of this annuity if a discount rate of 8 per cent applies.

If A = the amount received at the *end* of each year
P = present value of annuity = $A \times$ appropriate value from tables

At a discount rate of 8 per cent, and given that the payment is for ten years, the CIMA tables give a value of 6.71.

Therefore $P = £200 \times 6.71 = £1,342$

A perpetual annuity exists when the annual payments are made for ever. The formula for the present value of a perpetual annuity is simply:

$P = A/r$

Thus if, in the above example, the £200 payments will be made in perpetuity then the present value = 200/0.08 = £2,500.

8.3.8 Depreciation and amortization

This topic does not appear on the ICAEW or ICAS syllabuses

When a fixed asset, such as a machine, is purchased its cost can be apportioned over a number of accounting periods. In each year part of the cost (known as the depreciation) is written off against profits. Amortization refers to the similar process of writing off an expense, such as the purchase of a lease, where the lifetime of the asset is known for certain. Occasionally, however, the terms amortization and depreciation are used interchangeably. One way of allowing for depreciation is to set up a sinking fund.

8.3.9 Sinking funds

When it is known that a machine or piece of equipment will need to be replaced at some time in the future it is often a good idea to pay regular sums of money into a fund. This will accumulate, with interest, until it reaches the cost of replacing the asset at the appropriate time. Such a fund is known as a sinking fund. The payments made into the fund represent the depreciation of the existing asset.

Example

£2,000 is paid into a sinking fund at the end of each year. If the fund pays 12 per cent compound interest calculate the amount which will be in the fund at the end of four years.

Time of payment	Money paid into fund	Time money has spent in fund	Value of money at end of 4 years
End of year 1	£2,000	3 years	$2,000(1 + 0.12)^3 = £2,810$
End of year 2	£2,000	2 years	$2,000(1 + 0.12)^2 = £2,509$
End of year 3	£2,000	1 year	$2,000(1 + 0.12)^1 = £2,240$
End of year 4	£2,000	0 years	£2,000
		Value of fund at end of four years	= £9,559

Alternatively, a company may wish to know how much they should pay into a sinking fund annually to accumulate a given amount of money in the future.

Example

A firm anticipate that a replacement for their existing machine will need to be purchased in three years' time at a cost of £18,000. How much should they pay into a sinking fund at the end of each year if the fund pays 9 per cent interest?

Let x = amount paid into fund at end of each year.

Time of payment	Money paid into fund	Time money has spent in fund	Value of money at end of 3 years
End of year 1	£x	2 years	$x(1 + 0.09)^2 = 1.19x$
End of year 2	£x	1 year	$x(1 + 0.09)^1 = 1.09x$
End of year 3	£x	0 years	x
		Value of fund after three years	= $3.28x$
Therefore $3.28x = £18,000$		so $x = £5,488$	

9
Correlation and regression

The material of this chapter is relevant to the following syllabuses, except where otherwise indicated: CIMA, ICAEW, ICAI, ICAS, CACA(2.6), CIPFA, ICSA

9.1 Introduction

In Chapter 2 methods for analysing the characteristics of a single variable were discussed. However, sometimes the relationship between two variables is of interest. For example, are the sales of a product closely related to the amount spent on advertising it, or how are a factory's production costs related to its level of output?

Questions like these can often be answered by using correlation and regression analysis. Correlation is concerned with assessing how closely associated two variables are. Regression analysis attempts to describe the nature of the relationship between the variables and is mainly used to make predictions about one of the variables.

9.2 Correlation

Two variables which are closely associated are said to be correlated. Positive (or direct) correlation occurs where an increase in the value of one variable is associated with an increase in the value of the other variable. Thus one might expect advertising expenditure and sales to be positively correlated. Negative (or inverse) correlation occurs where an increase in the value of one of the variables is associated with a decrease in the value of the other. For example, warmer weather is associated with reduced heating bills.

9.2.1 Scattergraphs

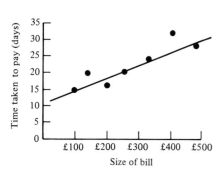

Figure 9.1

A scattergraph gives an initial impression of the strength of association between two variables. Figure 9.1 shows a scattergraph for the following data which relates to the size of electricity bills sent to seven randomly selected customers and the time the customers took to pay the bill.

Of course a sample of seven customers is rather small, and it will therefore only give a rough idea of the relationship between the two variables.

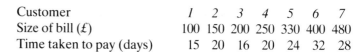

Customer	1	2	3	4	5	6	7
Size of bill (£)	100	150	200	250	330	400	480
Time taken to pay (days)	15	20	16	20	24	32	28

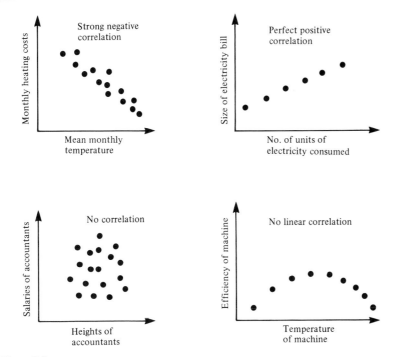

Figure 9.2

The scattergraph illustrates three important points. First, the greater the size of the bill then the longer it generally takes to pay. This clearly implies a positive correlation. Second, the scatter of points can be roughly summarized by a straight line. This is known as a linear association. Third, the points are scattered around, rather than upon, the line. Therefore the correlation is not perfect. It is not possible to predict exactly how long a customer will take to pay a bill of a particular size.

Figure 9.2 illustrates other patterns which can occur on scattergraphs.

9.2.2 Measuring correlation

A correlation coefficient gives a more precise indication of the strength of association between two variables. Two of the most widely used coefficients are

1 Pearson's product–moment coefficient (r)
2 Spearman's rank correlation coefficient (r').

Both measure correlation on a scale from -1 (perfect negative correlation) to $+1$ (perfect positive correlation) with 0 representing no correlation. Therefore a correlation of -0.9, for example, would represent a strong, negative correlation.

Note that both coefficients measure linear association. A strong curvilinear association may exist even if these coefficients suggest that there is no correlation.

A common mistake is to interpret a value of -0.9 as a very weak correlation. Remember that it is a strong correlation. The minus sign indicates that the correlation is negative.

Note that $(\Sigma x)^2$ is not the same as Σx^2.

The arithmetic here could be simplified by dividing all the x values by 10. Dividing all the numbers in either the x or y column by a constant, or subtracting a constant from all the numbers, has no effect on the final result.

(c) Pearson's product–moment coefficient

Pearson's product–moment coefficient can be calculated by using the following formula

$$r = \frac{n\,\Sigma xy - \Sigma x\,\Sigma y}{\sqrt{[n\,\Sigma x^2 - (\Sigma x)^2][n\,\Sigma y^2 - (\Sigma y)^2]}}$$

where one variable is labelled x, the other y and n = the number of pairs of observations.

The calculation for the electricity bill data is set out below.

x = the size of the bill y = the time taken to pay

x	y	xy	x^2	y^2
100	15	1,500	10,000	225
150	20	3,000	22,500	400
200	16	3,200	40,000	256
250	20	5,000	62,500	400
330	24	7,920	108,900	576
400	32	12,800	160,000	1,024
480	28	13,440	230,400	784
$\Sigma x = 1,910$	$\Sigma y = 155$	$\Sigma xy = 46,860$	$\Sigma x^2 = 634,300$	$\Sigma y^2 = 3,665$ $n = 7$

$$r = \frac{7(46,860) - 1,910(155)}{\sqrt{[7(634,300) - (1,910)^2][7(3,665) - (155)^2]}}$$

$$= 0.890$$

This clearly suggests a high positive correlation.

(b) Spearman's rank correlation coefficient

Spearman's rank correlation coefficient is applied to data which is rank form. It is useful where one or both of the variables cannot be easily measured but can be put into rank order. For example, it is difficult to measure variables like the skill of managers or the clarity of presentation of sets of accounts but one could rank the managers or the accounts from best to worst. Since the rank correlation formula is relatively simple to apply it is also used as an approximation for Pearson's coefficient.

The formula is

It is easy to misread this formula. Note that the 1 is not divided by anything.

$$r' = 1 - \frac{6\,\Sigma d^2}{n(n^2 - 1)}$$

where d = the difference between the pairs of ranks and n = the number of pairs of ranks.

For the electricity bill data the calculations are set out as follows. For each variable the largest number is given a rank of 1, the second largest 2 and so on. Note that two periods before payment tie for fourth place. Had the two times slightly differed they would have been fourth and fifth so they are both given a ranking which is the mean of 4 and 5, that is 4.5. The next highest time, of course, is ranked sixth.

If, for example, three numbers tied for second place, they would be given the mean of the ranks 2, 3 and 4, i.e. 3.

Size of bill	Ranks	Time to pay	Ranks	d	d²
100	7	15	7	0	0
150	6	20	4.5	1.5	2.25
200	5	16	6	−1	1
250	4	20	4.5	−0.5	0.25
330	3	24	3	0	0
400	2	32	1	1	1
480	1	28	2	−1	1
			$n = 7$ $\Sigma d =$	0	$\Sigma d^2 = 5.5$

Therefore $r' = 1 - \dfrac{6(5.5)}{7(49-1)} = 0.902$

which again suggests a high positive correlation.

9.2.3 Nonsense and spurious correlation

Correlation coefficients should always be interpreted with care. The existence of a high correlation between two variables does not prove that there is a causal link between them. There are two reasons for this.

First, the high correlation may be purely coincidental. For example daily sales of lemonade in Britain may be correlated with the daily number of road accidents in a South American city. This is called spurious correlation.

Second, the correlation may result from the effects of a third variable, a situation known as nonsense correlation. For example, a company may find that its weekly wage bill is correlated with the cost of travelling by rail. However, both variables are probably reflecting the level of inflation in the economy.

9.3 Regression analysis

Regression analysis is mainly used to predict the value of one variable (the dependent variable) from a knowledge of the value of the other variable (the independent variable). For example, sales may be predicted from a known level of advertising expenditure or production costs could be predicted for a known level of output.

Figure 9.3 shows a scattergraph for the data given below which relates to the production levels and production costs for five weeks in a factory.

Week	1	2	3	4	5
Number of units produced	2	4	6	8	10
Total production costs (£)	250	400	450	500	600

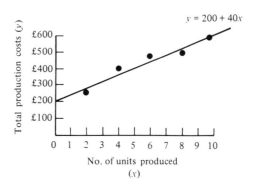

Figure 9.3

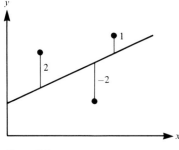

Figure 9.4

The scattergraph suggests that the relationship between production level and production costs can be approximately represented by a straight line. Regression analysis involves determining the equation of the line which most closely fits the scatter of points (the line of best fit).

To find this line the method of least squares is employed. Figure 9.4 illustrates the principle of this method. Here an attempt has been made to fit a line to a scattergraph of just three points. The vertical deviations of the points from the line are 2, -2 (if the point is below the line the deviation is negative) and 1. Therefore the sum of the squared vertical deviations is $2^2 + (-2)^2 + 1^2$ which equals 9. The line of best fit is the line which minimizes the sum of squared vertical deviations.

9.3.1 Finding the line of best fit

In Chapter 7 it was shown that a straight line can be represented by the equation

$$y = a + bx$$

where y is the dependent variable, x is the independent variable, a is the intercept of the line and b is the slope.

Note that b has to be calculated first, and the result is then substituted into the formula for a.

To find the slope and intercept of the line of best fit the following formulae are used

$$b = \frac{n\Sigma xy - \Sigma x \Sigma y}{n\Sigma x^2 - (\Sigma x)^2}$$

$$\text{and } a = \frac{\Sigma y}{n} - b\frac{\Sigma x}{n}$$

where n is the number of pairs of observations.

The application of these formulae to the production data is shown below. Note that, since the line is to be used to predict production costs, this is defined as the dependent variable (y).

x	y	xy	x^2	
2	250	500	4	
4	400	1,600	16	
6	450	2,700	36	
8	500	4,000	64	
10	600	6,000	100	
$\Sigma x = 30$	$\Sigma y = 2,200$	$\Sigma xy = 14,800$	$\Sigma x^2 = 220$	and $n = 5$

$$b = \frac{5(14,800) - 30(2,200)}{5(220) - (30)^2} = 40$$

$$a = \frac{2,200}{5} - \frac{40(30)}{5} = 200$$

Therefore the equation of the line of best fit, also known as the regression equation, is

$$y = 200 + 40x$$

This is the line superimposed onto the scattergraph in Figure 9.3. The equation can be interpreted as follows. If nothing is produced in a week, costs of around £200 will still be incurred. The intercept of the line therefore represents fixed costs. Also, for every extra unit produced the total cost will increase by an estimated £40. Thus the slope of the line represents variable costs.

9.3.2 Using the regression equation for prediction

To predict total production costs, the number of units produced is substituted for x in the equation and the value of y calculated. Thus if a production of five units is planned, production costs of $200 + 40(5) = £400$ would be predicted. There are, however, two types of prediction.

Interpolation involves predicting total production costs for a level of output which is within the observed range of output, that is between two and ten units. Extrapolation involves predicting production costs for a level of output which is outside the observed range. A prediction based on extrapolation should be treated with caution. Beyond the observed range of output there is no evidence that the linear relationship will still apply.

A regression equation should only be used to predict values of the variable which has been specified as the dependent variable. Therefore the above equation should not be used to predict the level of output which could be achieved for a given production cost. Instead a second equation would have to be calculated with the level of output now defined as the dependent variable (y) and production costs now defined as the independent variable (x).

Extrapolating too far beyond the range of observations can give nonsense results. A good example of this is given in Section 9.3.4.

9.3.3 The coefficient of determination

The square of Pearson's product–moment coefficient of correlation (r^2) is known as the coefficient of determination. This measure is useful in assessing how well a regression line fits a set of data. The coefficient of determination always lies between 0 and 1. If r^2 equals 1 a perfect fit has been achieved. For the data relating production costs to output level it can be shown that r equals 0.977 so that r^2 is 0.955, indicating that the fit is a good one.

A more precise interpretation of the coefficient of determination is that, when multiplied by 100, it shows the percentage of variation in one variable that can be explained by variations in the other variable. Thus 95.5 per cent of the variation in production costs over the five weeks has been explained by the different production levels. Other factors, such as variations in the quality of raw materials, will account for the remaining 4.5 per cent.

9.3.4 Forecasting future values of a time series

Regression analysis can be used to forecast the future values of a time series when it is thought that the underlying trend is linear. The sales of a company over a six-year period are shown below

Year	1981	1982	1983	1984	1985	1986
Sales (no. of units)	550	450	460	350	250	300

Sales will clearly be the dependent variable (y) and the years the independent variable (x). However, the calculations will be easier if x is set equal to 1 for 1981, 2 for 1982 and so on. It can be shown that the regression equation for this data is $y = 589.3 - 56x$. This suggests that sales have been falling by about fifty-six units per year. To forecast sales for 1987, $x = 7$ is substituted into the equation giving a forecast of about 197 units. However, the equation will give a forecast of about -27 units for 1991! This illustrates the dangers of 'nonsense extrapolation'.

9.4 Advanced topics in regression analysis

These topics only appear on the CACA(2.6) syllabus

9.4.1 The standard deviation of the regression

The standard deviation of the regression, S (sometimes called the residual standard error) is another measure of the closeness of the points on a scattergraph to the regression line. Its main use is in calculating confidence intervals for estimates based on regression lines.

$$S = \sqrt{\frac{\Sigma(y - \hat{y})^2}{n - 2}}$$

where $\hat{y}$ is the regression line's estimate of the y value, n is the number of points on the scattergraph.

Thus, for the production cost problem, where the regression equation was: $y = 200 + 40x$ the calculations are:

x	$\hat{y}$		y	$(y - \hat{y})^2$
2	$200 + 40(2)$	$= 280$	250	900
4	$200 + 40(4)$	$= 360$	400	1,600
6	$200 + 40(6)$	$= 440$	450	100
8	$200 + 40(8)$	$= 520$	500	400
10	$200 + 40(10)$	$= 600$	600	0

therefore $\Sigma(y - \hat{y})^2 = 3,000$ and $n = 5$

so that $S = \sqrt{\dfrac{3,000}{5 - 2}} = 31.6$

A value of zero would indicate a perfect fit.

9.4.2 Confidence intervals for regression

Predictions made from regression lines can be expressed as confidence intervals. A 95 per cent confidence interval for the average value of y, for a given value of x, can be calculated using the following formula

$$\hat{y} \pm t_{0.025} S \sqrt{\frac{1}{n} + \frac{(x - \bar{x})^2}{\Sigma(x - \bar{x})^2}}$$

where $\hat{y}$ is the regression line's prediction of y, S is the standard deviation of the regression, n is the number of points on the scattergraph and $t_{0.025}$ is found in the t tables (see Section 6.2.6) where the number of degrees of freedom is $n - 2$.

Example

For the production cost problem, derive a 95 per cent confidence interval for the average level of production costs for weeks when eight units are produced.

The regression equation is $y = 200 + 40x$.
Therefore, if $x = 8$, $\hat{y} = 200 + 40(8) = £520$.
$S = 31.6$ (see last section), $n = 5$, $\bar{x} = $ mean value of the $xs = 6$ so that $\Sigma(x - \bar{x})^2 = 40$.
There are $5 - 2 = 3$ degrees of freedom so $t_{0.025} = 3.18$.
Therefore the 95 per cent confidence interval

$$= 520 \pm 3.18(31.6) \sqrt{\frac{1}{5} + \frac{(8 - 6)^2}{40}} = £465 \text{ to } £575$$

That is, the management can be 95 per cent confident that, for weeks when eight units are produced, the average level of production costs will fall within this interval.

9.4.3 Multiple regression

Sometimes a better prediction can be made when there is more than one independent variable in a regression model. For example, the monthly sales of a product could be predicted not only from the amount spent on advertising it, but also from its price and the price of a competing product. A typical model might be:

$$y = 200 + 3x_1 - 4x_2 + 0.5x_3$$

where y is the sales of the product, x_1 is advertising expenditure, x_2 is the price of the product and x_3 is the price of the competing product.

The *multiple correlation coefficient* (R) can be used to assess how well a multiple regression model fits the data from which it is derived. In the above example it would measure the correlation between the sales predicted by the model ($\hat{y}$) and the actual sales (y). The *multiple coefficient of determination* (R^2) would give the proportion of variation in sales which could be explained by variations in the three independent variables.

9.4.4 Non-linear regression

In some problems the relationship between the variables can be modelled by a curve. Two curvilinear functions are often used.

(a) Exponential functions

In exponential functions the relationship between the two variables is represented by: $y = ab^x$.

Situations where there is a constant percentage rate of growth can be modelled by this equation (e.g. a firm's sales may be growing at 10 per cent per year). To find the curve which best fits a set of data it is necessary to take logarithms of both sides of the equation so that:

$$\log y = \log a + (\log b)x$$

Comparing this with: $y = a + bx$

it can be seen that the result is a linear function. This means that the regression formulae in Section 9.3.1 can be used with $\log y$ replacing y, $\log a$ replacing a, and $\log b$ replacing b to give:

$$\log b = \frac{n\Sigma x(\log y) - \Sigma x \Sigma \log y}{n\Sigma x^2 - (\Sigma x)^2}$$

$$\text{and } \log a = \frac{\Sigma \log y}{n} - \log b \frac{\Sigma x}{n}$$

Example
The sales of a company for a five year period are given below. Fit a curve of the form $y = ab^x$ to the data and use it to predict sales in year 5.

(x) Year	1	2	3	4
(y) Sales (no. of units)	62	82	125	162

Using logs to the base 10:

x	y	$log\,y$	$x\,log\,y$	x^2
1	62	1.79	1.79	1
2	82	1.91	3.82	4
3	125	2.10	6.30	9
4	162	2.21	8.84	16

$$\Sigma x = 10 \qquad \Sigma \log y = 8.01 \qquad \Sigma x \log y = 20.75 \quad \Sigma x^2 = 30 \text{ and } n = 4$$

$$\text{Thus } \log b = \frac{4(20.75) - 10(8.01)}{4(30) - (10)^2} = 0.145$$

$$\text{so } b = \text{antilog } (0.145) = 1.4$$

$$\text{and } \log a = \frac{8.01}{4} - \frac{0.145(10)}{4} = 1.64$$

$$\text{so } a = \text{antilog } (1.64) = 43.65$$

Therefore the equation is $y = 43.65(1.4)^x$ so that the sales forecast for year 5 is: $y = 43.65(1.4)^5 = 235$ units.

(b) Geometric functions

Geometric functions have the equation: $y = ax^b$

Learning curves are one well-known application of this function. On taking logarithms, the equation becomes: $\log y = \log a + b \, \log x$. The regression equation can thus be obtained by using the regression formulae with $\log y$ replacing y, $\log a$ replacing a and $\log x$ replacing x. Thus:

$$b = \frac{n\Sigma(\log x)(\log y) - \Sigma \log x \Sigma \log y}{n\Sigma(\log x)^2 - (\Sigma \log x)^2}$$

$$\text{and } \log a = \frac{\Sigma \log y}{n} - \frac{b\Sigma \log x}{n}$$

10
Linear programming

The material in this chapter is relevant to the following syllabuses, except where otherwise indicated: CIMA, ICAEW, CACA(2.6), CIPFA, ICSA

10.1 Introduction

The last three chapters of this book are concerned with operational research (OR). It is extremely difficult to give a concise definition of OR, but the subject is essentially concerned with the adoption of a scientific approach to decision problems which occur in industrial, commercial, governmental or military systems where resources like labour, money, materials and machines need to be managed. This chapter outlines the principles of an important OR technique, namely linear programming, and looks at two related techniques, the transportation and assignments methods.

10.2 Linear programming

10.2.1 Formulating models

Linear programming is a method which can be applied to decision problems where it is necessary to make optimal use of limited resources (e.g. limited time, cash, raw materials or storage space). The problem is represented by a mathematical model and formulating the model involves the following steps.

1 *Identify the decision variables*   These are the variables whose value the decision maker wishes to determine. For example, the decision variables might be the production levels of two products or the amount of money to be allocated to three investment projects.
2 *Identify the objective* (e.g. maximizing contribution or minimizing costs) and formulate it as an equation involving the decision variables (this is known as the objective function).
3 *Identify the constraints* and formulate them as mathematical expressions involving the decision variables.

Example

Two products, A and B, are manufactured on the same equipment at a chemical plant. The plant can operate for only forty-eight hours per week and each unit of A takes two hours to produce while each unit of B takes three hours. Furthermore, each unit of A produced requires 200 kg of a raw material while each unit of B requires 100 kg of the same material. However, because of supply problems, only 2,000 kg of the material is available per week. Also, for marketing reasons, at least six units of B must be produced per week. If each unit of A produced earns a contribution of £400 while each unit of B earns £500 how many units of A and B should be produced per week to maximize contribution?

The linear programming model of the problem is formulated as follows.
1 Identify the decision variables
    Let a = the number of units of product A produced per week and b = the number of units of product B produced per week
2 Formulate the objective function
    Maximize C = $400a + 500b$ (where C = weekly contribution in £s)
3 Identify and formulate the constraints

Time constraint:	$2a + \quad 3b \le$	48 (hours)
Raw materials constraint:	$200a + 100b \le 2{,}000$	(kg)
Marketing constraint:	$b \ge$	6 (units)

    Also, since it is not possible to produce a negative number of units: $a > 0$ and $b \ge 0$ (these are called non-negativity constraints)

≤ means 'less than or equal to' while ≥ means 'greater than or equal to'.

Really b≥0 is redundant in this case since we have already said that b≥6.

10.2.2 Solving linear programming problems using the graphical method

Because there are only two decision variables in the above problem the graphical solution method can be used (see Figure 10.1). Each axis of the graph represents the weekly output of one of the products. The constraints are then plotted on the graph as if they are equations (i.e. with an equals

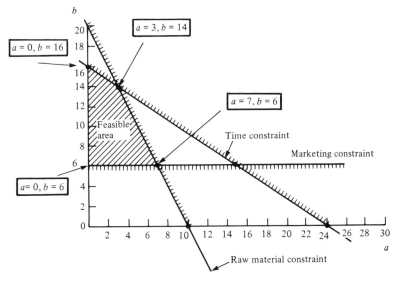

Figure 10.1

A simple way of plotting a constraint is to find the value of each variable when the other variable is 0. Thus for the raw materials constraint if $a = 0$, $0 + 100b = 2,000$ so that $b = 20$ (i.e. the line passes through the point $a = 0, b = 20$). If $b = 0$, $200a + 0 = 2,000$, so that $a = 10$ (i.e. the line also passes through the point $a = 10, b = 0$).

sign replacing the inequality). Note that the infeasible sides of the constraints have been marked. For example, the combined outputs of A and B must not require more than 2,000 kg of raw material so the upper side of this constraint has been marked. The only combinations of output which satisfy all the constraints lie in, or on the edge of, the shaded area. This is therefore known as the feasible area.

It can be proved that the optimal combination of output must lie at one of the corners of the feasible area. (Occasionally it may lie at two adjacent corners which means that all the points on the line joining the two corners are also optimal.) The combinations of output at the corners of the feasible area are given below together with the contribution generated by these levels of output. Note that simultaneous equations were used to determine the points where lines intersected.

a	b	$C = 400a + 500b$
0	6	£3,000
0	16	£8,000
3	14	£8,200
7	6	£5,800

Clearly, the optimum solution is to produce three units of A and fourteen of B per week to earn a contribution of £8,200.

10.2.3 Shadow prices

Sensitivity analysis is often applied to linear programming models to examine the effects of changes in the data which has been used in the model. An important result which can be determined by sensitivity analysis is a dual or shadow price. This is the change in the value of the objective function which would result if the right-hand side of a constraint was changed by one unit. In the context of the above model a shadow price is the increase in contribution which would result if one more unit of a resource became available (or if the marketing constraint was relaxed by one unit). For example, if the plant could be run for forty-nine hours (instead of forty-eight) the time constraint would become: $2a + 3b \leq 49$. If the modified model was then solved the optimum outputs would be 2.75 units of A (assuming that this output is possible) and 14.5 units of B which would give a contribution of £8,350. Thus the extra hour would increase contribution by £150 per week and this is therefore the shadow, or dual, price of production time. Similarly, an hour's reduction in time to forty-seven hours would reduce contribution by £150.

10.2.4 Limitations of linear programming

For linear programming to be applied to a problem all the constraints and the objective function must be linear functions, and, strictly speaking, it must be possible for the decision variables to have fractional values (e.g.

2.75 units of product A above). The method also assumes that all the information in the model (e.g. the availability of resources) is known with certainty. Moreover, the method can only be used where one objective has to be achieved. In many problems the decision maker is faced with a number of, often conflicting, objectives.

The remaining topics in this chapter are relevant only to the CACA(2.6) and CIPFA syllabuses

10.2.5 Linear programming problems with more than two decision variables

The graphical method cannot be used to solve linear programming problems where there are more than two decision variables. However, problems with any number of decision variables can be solved by using a technique known as the simplex method. This involves some rather tedious calculations and, in practice, most problems are solved by using computer packages. However, many packages give the solution in the form of a Simplex tableau and it is important to be able to interpret this.

Consider the following linear programming model. This represents the problem of finding the weekly production levels of three products which will maximize contribution. The output of the products is constrained by the availability of a raw material and by the person-hours available in the machine and packaging centres.

Maximize $C = 4x_1 + 2x_2 + 6x_3$ subject to

$$x_1 + 2x_2 + 2x_3 \le 450 \text{ (raw materials constraint)}$$
$$3x_1 + x_3 \le 480 \text{ (time in machine centre constraint)}$$
$$x_1 + 4x_2 \le 200 \text{ (time in packaging centre constraint)}$$
$$x_1, x_2, x_3 \ge 0$$

where x_1, x_2 and x_3 = the weekly outputs of products 1, 2 and 3 respectively.

Initially, the simplex method allocates slack variables to the constraints, as shown below. These remove the inequalities from the constraints and they represent the spare resources which exist at a given level of output. For example, if, in a given week, ten units of each product were produced this would use up $10 + 2(10) + 2(10) = 50$ units of the raw material. Thus S_1, the number of units of the raw material which were spare, would equal 400.

Maximize $C = 4x_1 + 2x_2 + 6x_3$ subject to

$$x_1 + 2x_2 + 2x_3 + S_1 = 450$$
$$3x_1 + x_3 + S_2 = 480$$
$$x_1 + 4x_2 + S_3 = 200$$

where S_1 = the number of spare units of the raw material; S_2 = the number of spare person hours in the machine centre and S_3 = the number of spare person-hours in the packaging centre.

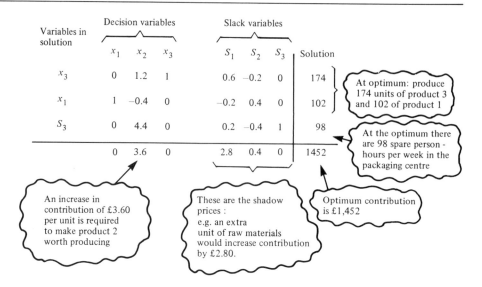

Figure 10.2

Note that sometimes the optimal simplex tableau is given with a bottom row consisting of negative numbers. If this is the case, the minus signs should be ignored when interpreting the tableau.

Figure 10.2 shows the simplex tableau for the optimal solution to this problem and illustrates how this solution can be interpreted. The main information is obtained from the end column and the bottom row of the tableau and can be grouped under four headings.

1   *The optimum solution to the problem and the resources which are spare at this solution*   Note that the 'Variables in solution' column is often omitted. In this case, the optimal solution can be found by identifying the columns which contain only a single 1 and zeroes (e.g. the x_3 column). If each of the ones is replaced by its column heading, it can be seen that row 1 is the x_3 row, row 2 is the x_1 row and row 3 is the S_3 row. The value of the variables is then found by looking at the numbers in the end column. Thus $x_3 = 174$, $x_1 = 102$ and $S_3 = 98$. Any variables not in the solution have a value of zero. Therefore, x_2, S_1 and S_2 all equal 0 at the optimal solution.

2   *The optimal weekly contribution*   This is found at the bottom right-hand corner of the tableau and equals £1,452.

3   *The shadow prices of the resources*   These are found in the bottom row under the slack variable columns. Note that a person-hour in the packaging centre has a shadow price of £0 because there are already ninety-eight person-hours spare so an increase in the number of hours available will not increase the total contribution.

4   *The increases required in the contributions per unit of the products to make them worth producing*   These are found in the bottom row under the decision variable columns. Note that no increase is required in the contributions of products 1 and 3 since the solution shows that they are already worth producing.

10.3 Transportation linear programming

Consider the following problem. A company has three factories located in towns A, B and C which supply four warehouses, located in towns W, X, Y and Z with a product. The weekly requirements of the warehouses and the weekly production capacities of the factories are shown in Table 10.1 together with the costs of transporting a unit of the product from the factories to the warehouses. The objective of the company is to determine the number of units which should be sent from each factory to each warehouse in order to minimize total transport costs.

Table 10.1

Factories	Warehouses				Production capacity (units)
	W	X	Y	Z	
A	£16	£18	£12	£5	17
B	£12	£22	£10	£20	20
C	£6	£16	£14	£18	18
Requirements (units)	15	16	11	13	

This problem could be formulated as a linear programming model. However, a more efficient solution procedure, known as the transportation method, has been developed for this type of problem. The method is also applicable to some problems which have nothing to do with transportation, but which have a similar structure (e.g. certain production planning problems). The transportation method involves the following steps.

Step 1   Determine a first feasible solution

This is a solution where the requirements of the warehouses will be met, but it may not be the solution with the minimum transport cost.

It can be obtained in a number of ways. The method used below is called the least-cost method.

1   First, as many units as possible are allocated to the cheapest route AZ. The maximum that can be allocated here without exceeding the row or column totals is thirteen units. Z's requirements are now met so zeroes are allocated to the other routes in this column.

2   However, A still has four units to supply. The cheapest remaining route from A is AY so four units are allocated here and zeroes allocated elsewhere in row A.

3   Now Y still requires seven units and the cheapest way to supply these is from B. Y's requirements are now met so a zero is allocated to the remaining route in this column.

The method proceeds in the same way, alternately completing the rows and columns, until the amount transported along every route has been determined (see Table 10.2). Now the solution needs to be tested to see if it minimizes costs.

Table 10.2

Factories	Warehouses				Production capacity (units)
	W	X	Y	Z	
A	0 £16	+K 0 £18	−K 4 £12	13 £5	17
B	−K 13 £12	0 £22	+K 7 £10	0 £20	20
C	+K 2 £6	−K 16 £16	0 £14	0 £18	18
Requirements (units)	15	16	11	13	

Step 2   Calculate the shadow costs of each used route

For each used route the transport cost is split into an imaginary despatch cost at the factory and an imaginary reception cost at the warehouse. Thus, the following equations can be derived where, for example, A represents the despatch cost at factory A and Y the reception cost at warehouse Y. $A + Y = 12$, $A + Z = 5$, $B + W = 12$, $B + Y = 10$, $C + W = 6$, and $C + X = 16$.

These can be solved by giving A an arbitrary value of 0. Thus $Y = £12$, $Z = £5$, $B = -£2$, $W = £14$, $C = -£8$ and $X = £24$.

Step 3   For each unused route, calculate the effect of bringing the route into use

This can be achieved by calculating

Cost of the unused route − sum of the shadow costs for warehouse and factory on this route

Thus: Cost of $AW - (A + W) = 16 - (0 + 14) = +£2$. (This means that using this route would increase costs by £2 for every unit sent along the route.) Similarly

$$
\begin{aligned}
\text{Cost of AX} - (A+X) &= 18 - (\ 0+24) = -£6\\
\text{Cost of BX} - (B+X) &= 22 - (-2+24) = \ \ £0\\
\text{Cost of BZ} - (B+Z) &= 20 - (-2+\ 5) = £17\\
\text{Cost of CY} - (C+Y) &= 14 - (-8+12) = £10\\
\text{Cost of CZ} - (C+Z) &= 18 - (-8+\ 5) = £21
\end{aligned}
$$

It can be seen that £6 can be saved for each unit sent along route AX and so the current solution is not optimal.

Step 4   If the solution is not optimal, transfer as many units as possible to the route which will give the biggest saving

Since only route AX gives a saving, as many units as possible are transferred to this route. This is achieved by using the 'stepping stone' method (see the last tableau).

Let k = the number of units to be transferred to AX. Therefore $+k$ is written in this square representing this route. This means that k units need to be subtracted from another route in the X column. This route can only be CX (subtracting from BX would lead to a negative number of units on this route). Now k units need to be added to another route in the C row. This route can only be CW (or another unused route is being brought into the solution). Therefore k units must be subtracted from BW, and so on. This process is continued until every row or column which has been 'disturbed' contains both a $+k$ and a $-k$.

The largest value which k can have, without making the number of units carried along a route negative, is 4. The revised solution is given in Table 10.3.

If steps 2 and 3 are repeated, it will be seen that this is an optimum solution, since the cost of bringing any of the unused routes into the solution can only

The three key points to remember when using the stepping stone method are: (a) the totals of the rows and columns must remain the same; (b) the number of units sent along a route cannot be negative; (c) only one unused route must be brought into the solution at a time.

Table 10.3

Factories	Warehouses W	X	Y	Z	Production capacity (units)
A	0 £16	4 £18	0 £12	13 £5	17
B	9 £12	0 £22	11 £10	0 £20	20
C	6 £6	12 £16	0 £14	0 £18	18
Requirements	15	16	11	13	

increase costs. (There is one exception to this. Bringing route BX into the solution would make no difference to the total costs and therefore there is more than one optimal solution to this problem.) The total weekly transport cost at this solution is found by multiplying the number of units carried along each route by the cost of that route, and adding the results. The total cost here is £583.

10.4 Extensions of the transport method

1   If the total number of units required by the warehouses does not equal the total capacity of the factories, a dummy row or column has to be introduced into the matrix to take up the slack. The costs of the routes in the dummy row or column are zero.
2   If it is not possible to use a route (e.g. because a road link is temporarily closed), the cost of the route should be made extremely high (e.g. £9,999) to ensure that the method does not allocate units to it.
3   If the monetary values on the tableau represent the profits of using a route rather than costs, and the objective is to maximize profit, the method can be 'tricked' into allocating to the most profitable routes by initially subtracting every profit figure from the highest profit figure.
4   When the number of used routes is less than the number of factories + the number of warehouses − 1 (i.e. 6 in the above example) then it will not be possible to calculate all the shadow costs, a situation which is referred to as degeneracy. This problem is solved by simply treating one or more unused routes as used routes in steps 2 and 3.

10.5 The assignment method

Suppose that a manager has to allocate four engineers to four repair jobs in factories which are located in different parts of the country. The distance of the factories from each engineer's home is shown below. The manager wants to find the allocation which will minimize the total distance travelled by the engineers.

	Factory (distances to nearest 10 miles)			
	A	B	C	D
Engineer				
P	120	80	250	200
Q	300	200	60	160
R	240	80	80	180
S	40	280	220	90

The assignment problem is really a special form of the transportation problem. The requirements at each destination are one unit (or person), and one unit (or person) is available at each source. However, the problem is always degenerate.

The best allocation can be made by using the assignment method which involves the following steps.

Step 1

Subtract the smallest element in each row from every element in that row. This gives

	Factory			
	A	B	C	D
Engineer				
P	40	0	170	120
Q	240	140	0	100
R	160	0	0	100
S	0	240	180	50

Step 2

Subtract the smallest element in each column from every element in that column. This gives

	Factory			
	A	B	C	D
Engineer				
P	40	0	170	70
Q	240	140	0	50
R	160	0	0	50
S	0	240	180	0

Step 3

By drawing horizontal or vertical lines across rows or columns, attempt to 'cover up' all the zeroes with a minimum number of lines. If this can be achieved in less than *n* lines (where *n* is the number of rows or columns of the matrix) an optimal solution is not yet possible. Otherwise proceed to step 5.

	Factory			
	A	B	C	D
Engineer				
P	40	0	170	70
Q	240	140	0	50
R	160	0	0	50
S	0	240	180	0

Three lines can cover all the zeroes here, so an optimal allocation is not yet possible.

Step 4

Subtract the smallest uncovered element from all the other uncovered elements, and add it to the elements crossed by two lines. Leave the other elements as they were. Then repeat step 3.

	Factory			
	A	B	C	D
Engineer				
P	0 ✓	0 ×	170	30
Q	200	140	0 ✓	10
R	120	0 ✓	0 ×	10
S	0 ×	280	220	0 ✓

Four lines are needed to cover all the zeroes. Therefore an optimum allocation can now be made. (If this were not the case this step would have to be repeated.)

Step 5

Make the optimum allocation. This is achieved by using cells with a zero element. First look for a row or column containing one zero (e.g. column D). Thus engineer S goes to factory D. This means that the zero in SA cannot be used and it is therefore crossed out. Thus P must go to A which rules out the zero in PB. Therefore R goes to B which rules out the zero in RC. This means that Q goes to C.

From the original matrix it will be seen that this allocation gives a total distance to be travelled of 350 miles.

Note that if the assignment matrix is not square, a dummy row or column must be introduced with zero elements, as in the transport method. Other problems can be dealt with in the ways that were suggested for the transport method in Section 10.4.

11
Stock control, queuing and simulation

The material of this chapter is relevant to the following syllabuses, except where otherwise indicated: CIMA, ICAS, CACA(1.5), CACA(2.6), CIPFA

11.1 Introduction

This chapter considers the application of models to stock control, and queuing problems. The use of simulation models is also considered.

11.2 Stock control models

11.2.1 Basic ideas

The following terms are used in stock control.

1 The *lead time* is the time between placing an order for goods and actually receiving them.
2 *Holding costs* are the costs incurred by holding goods in stock. They include storage and insurance costs and the cost of having capital tied up in stock.
3 *Ordering costs* are the costs which are incurred each time an order is placed. They include delivery costs and the administrative costs of placing an order.
4 A *stock-out* occurs when an item is required which is not in stock.
5 *Stock-out costs* are the costs incurred because an item is not in stock (e.g. the costs of having to place an emergency order).
6 *Safety stock* is the stock held to cater for above-average demand during the lead time.

Operational Research models can be used in stock control to answer two questions

1 How much stock should be ordered at a time? Economic order quantity models are used to answer this question.
2 When should an order be placed? Two systems are commonly in use: the re-order level system and the periodic review system.

Note the definition of a stock-out carefully. If 100 units of a good were available for a given month's trading and exactly 100 units were demanded by customers then we would have no stock left at the end of the month. However, a stock-out would not have occurred because there was no demand for the good which could not be met.

11.2.2 The economic order quantity (EOQ) model

The assumptions of this model are

1 Demand for the item of stock is constant.
2 The lead time is either zero or is known for certain.

3 No stock-outs are allowed.
4 Holding costs are proportional to the amount of stock held (i.e. if the stock level doubles so do the holding costs).
5 The cost of placing an order is constant (i.e. it costs as much to place a large order as a small order).

The EOQ model can be applied as an approximation where these rigid conditions are not exactly met.

If these conditions are fulfilled, and if Q units are ordered at a time, the average stock level will be $Q/2$ (as shown in Figure 11.1). Thus, large orders mean that more stock has to be held so that stockholding costs will be high. However, large orders will also lead to low ordering costs because fewer orders need to be placed. The economic order quantity is the size of order which will minimize the sum of stockholding and ordering costs. The formula for the EOQ is

$$EOQ = \sqrt{\frac{2C_o D}{C_h}}$$

If demand is not constant, average demand could be used instead in the formula.

where D is the annual demand for the stock item, C_o is the cost of placing an order, C_h is the cost of holding one unit of stock for a year.

Example

An electronic component is used at a constant rate of four per week for fifty weeks a year. Each unit is bought at a price of £4 and the cost of holding a component for a year is 20 per cent of its price. A charge of £5 is incurred each time an order is placed. What is the EOQ?

$D = 4 \times 50 = 200$, $C_h = 20\%$ of £4 $= £0.8$ and $C_o = £5$ therefore

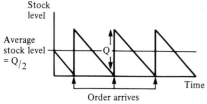

Figure 11.1

$$EOQ = \sqrt{\frac{2 \times 5 \times 200}{0.8}} = 50 \text{ units}$$

Therefore fifty units should be ordered at a time. Note also that, since the annual demand is 200, four orders (i.e. 200/50) will be placed per year. The average stock held will be 50/2 = 25 units.

11.2.3 Derivation of the EOQ formula

This topic does not appear on CACA(2.6) syllabus

Annual ordering costs = cost of placing order × no. of orders placed per year, but the number of orders placed per year is

$$\frac{\text{Annual demand}}{\text{size of order}}$$

so

Annual ordering costs = cost of placing order $\times \dfrac{\text{annual demand}}{\text{size of order}}$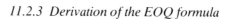

$$= C_o \times \frac{D}{Q}$$

Annual holding costs = average stock level × cost of holding a unit for a year

$$= \frac{Q}{2} \times C_h$$

Therefore

$$T = \text{annual ordering} + \text{holding costs} = \frac{C_o D}{Q} + \frac{Q C_h}{2}$$

To find the value of Q which minimizes T the above function can be differentiated with respect to Q, and the derivative equated to zero. Thus

$$\frac{dT}{dQ} = \frac{-C_o D}{Q^2} + \frac{C_h}{2} = 0 \qquad \text{therefore} \frac{C_h}{2} = \frac{2 C_o D}{Q^2}$$

$$\text{which leads to } Q = \sqrt{\frac{2 C_o D}{C_h}}$$

This is a minimum point because

$$\frac{d^2 T}{dQ^2} = \frac{2 C_o D}{Q^3}$$

which must be positive since Q is always positive.

11.2.4 Discounts for ordering in bulk

This topic does not appear on CACA(1.5) syllabus

If a supplier offers a bulk purchase discount it may be worth ordering more than the EOQ. Suppose that in the last example the supplier of electronic components offers a 10 per cent discount on the price of a component on orders of 200 or more units. Is it worth accepting the offer?

Because the discount is offered on the purchase price, the annual purchasing costs now also need to be considered so that

Total annual costs
= annual purchasing costs + annual ordering costs
+ annual holding costs
= (annual demand × price) + (no. of orders × cost of order)
+ (average stock × cost of holding each unit)

At the EOQ, total annual costs = $(200 \times £4) + (4 \times £5) + (25 \times £0.8) =$ £840.

Note that just enough units are ordered to obtain the discount.

If 200 units are ordered the discount price per unit will be 90 per cent of £4 = £3.60. Thus the cost of holding a unit for a year = 20 per cent of £3.60 = £0.72.

The average stock held will be 200/2 = 100 units and only one order will be placed per year. Therefore

Total annual costs = $(£3.60 \times 200) + (1 \times £5) + (100 \times £0.72)$
= £797

So it is worth accepting the offer and ordering 200 units at a time.

11.2.5 The economic batch quantity (EBQ) model

This topic does not appear on the CACA(1.5) syllabus

The economic batch quantity (EBQ) model can be used where the stock-holder also manufactures the product. When necessary a batch of units is produced and stocks are *gradually* replenished during the production run. Each time a production run is started, set-up costs (e.g. the costs of cleaning or adjusting machines for the new production run) are incurred. If a large number of items is produced in a batch then the machines will have to be set up less often so that annual set-up costs will be lower. However, large batch sizes mean that more stock will have to be held. The economic batch quantity is the batch size which will minimize the sum of annual set-up costs and holding costs.

It can be shown that

$$EBQ = \sqrt{\frac{2C_oD}{C_h\left(1-\frac{D}{R}\right)}}$$

where C_o is now the cost of setting up a production run, C_h is the stock-holding cost (as before), D is the annual demand (as before) and R is the rate of production (this is the number of items which could be produced in a year if production took place without interruption).

Example

A company uses 6,000 units of a component per year. If necessary, the component can be produced at a rate of 200 per week (assume that there are fifty weeks in a year). Setting up a production run costs £400 and holding a unit in stock for a year costs £12. What is the EBQ?

$D = 6,000$, $C_h = £12$, $C_o = £400$ and $R = 200 \times 50 = 10,000$ therefore

$$EBQ = \sqrt{\frac{2 \times 400 \times 6,000}{12\left(1-\frac{6,000}{10,000}\right)}} = 1,000 \text{ units}$$

Note that this implies that there will be $6,000/1,000 = 6$ production runs per year, and, since 200 units can be produced per week, a production run will take $1,000/200 = 5$ weeks to complete.

11.2.6 The re-order level system

This topic does not appear on the CACA(1.5) syllabus

The re-order level (ROL) (or two-bin) system can be used to determine when orders for new stock should be placed. Orders are placed as soon as stocks have been depleted to a certain level: the re-order level.

If the length of the lead time and the demand for a product are known for certain then the re-order level = stock which will be needed in the lead time. Thus, if the demand for a product is always two units per day and, when an order is placed, it always takes six days to arrive then the ROL = $6 \times 2 = 12$ units (i.e. as soon as 12 units are left in stock an order should be placed).

If the demand for the product and/or the length of the lead time is uncertain then the re-order level = average usage of stock in lead time + safety stock. The level of safety stock, and hence the ROL, can be designed to ensure that stock-outs only occur in x per cent of lead times.

At the time of placing the order we want to ensure that we have enough stock to 'keep us going' until the order is delivered.

Example

The weekly demand for a product is normally distributed with a mean of twenty units and a standard deviation of four units and each week's demand is independent of other weeks. When an order is placed the lead time is three weeks and the company aim to run out of stock in no more than 5 per cent of lead times. What should the ROL be?

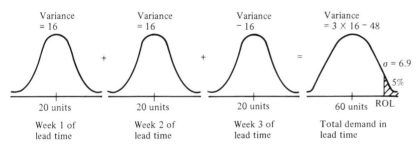

Figure 11.2

Since the lead time is three weeks, the distribution of total demand in the lead time can be found by adding the three normal distributions of weekly demand as shown in Figure 11.2. The ROL is the level of stock which only has a 0.05 probability of being exceeded. Normal distribution tables show that the Z value of the ROL is about 1.65, so that

$$ROL = 60 + 1.65 \times SD$$

i.e. $ROL = 60 + 1.65 \times 6.9$
$= 71$ units (approx.)

(i.e. a safety stock of eleven units, since the average demand in the lead time is sixty units).

11.2.7 The periodic review system

This topic does not appear on the CACA(1.5) syllabus

In the periodic review system orders are placed at fixed intervals of time (e.g. every five weeks). However, the amount of stock ordered will vary from one order to the next. This is because the amount ordered is designed to bring stocks up to some predetermined level, so it will depend on the level of stocks at the time of placing the order.

The periodic review system has administrative advantages over the ROL system because the timing of orders can be planned. However, it can mean that more stock has to be held because the system is less sensitive to variations in demand.

11.3 Queuing theory

This topic appears only on the CACA(2.6) and CIPFA syllabuses

11.3.1 Introduction

A large number of OR models have been developed to represent queuing problems. Often these problems involve a decision about the level of a service facility (e.g. how many cash desks to operate in a bank).

The following terms are used in queuing theory
1   The *system* refers to the customers in the queue awaiting service plus the customers actually being served.

2   The *traffic intensity* $= \dfrac{\text{the mean arrival rate of customers}}{\text{mean rate at which they can be served}}$

3   The *queue discipline* refers to the order in which customers are served and other features of their behaviour such as queue switching.

(M/M/1) is the standard notion for queue with a Poisson arrival rate, Poisson service rate, and one service channel.

11.3.2 The simple queue (M/M/1) model

The assumptions of this model are
1   The number of customers arriving per period follows a Poisson distribution.
2   The number of customers who can be served per time period also follows a Poisson distribution.
3   The queue discipline is first come first served.
4   There is an infinite population of potential customers and no limit to the queue's length.
5   Customers are not allowed to leave the queue before being served.
6   There is only one service channel.
7   The traffic intensity is less than one (i.e. the average number of customers arriving per period is less than the average number who can be served).

If the number of arrivals follows a Poisson distribution then the time interval between arrivals follows a distribution called the negative exponential distribution. Similarly, if the number of customers who can be served per time period follows a Poisson distribution then the service times follow a negative exponential distribution.

If these assumptions apply then the following formulae can be used where λ (lambda) is the mean arrival rate of customers per period and μ (mu) is the mean number of customers who can be served per period.

Be very careful not to confuse service times with service rates e.g. if a customer takes 30 minutes to serve, on average, then the average service rate (μ) is two per hour. The same is true of inter-arrival times. If the average interval between customer arrivals is 15 minutes then the average arrival rate (λ) is four per hour.

1   Mean time a customer spends in the queue $= \dfrac{\lambda}{\mu(\mu - \lambda)}$

2   Mean time a customer spends in the system $= \dfrac{1}{\mu - \lambda}$

3   Mean number of customers in the queue $= \dfrac{\lambda^2}{\mu(\mu - \lambda)}$

4   Mean number of customers in the system $= \dfrac{\lambda}{\mu - \lambda}$

Example

If customers arrive in a simple queue at a mean rate of seven per hour and can be served at a mean rate of ten per hour then calculate the mean time customers will spend in (a) the queue and (b) the system.

$\lambda = 7$ and $\mu = 10$

$$\text{Mean time customer spends in queue} = \frac{7}{10(10 - 7)}$$

$$= \frac{7}{30} \text{ hours} = 14 \text{ minutes}$$

$$\text{Mean time a customer spends in the system} = \frac{1}{10 - 7}$$

$$= \frac{1}{3} \text{ hours} = 20 \text{ minutes}$$

(i.e. 14 minutes in the queue plus 6 minutes being served).

11.3.3   Using queuing theory for decision making

Example

A company has a very large number of machines which break down at an average rate of four per hour. The machines are repaired in the order in which they break down, but production valued at £50 is lost for each hour a machine is out of action. Currently, the company employs eight mechanics who, working as a team, can repair a machine in an average time of twelve minutes. Each mechanic is paid £4 per hour. It is estimated that, if a ninth mechanic was employed repairs could be completed in an average time of ten minutes. Assuming that the (M/M/1) queuing model applies, is it worth employing the extra mechanic?

For the eight-mechanic team:

$\lambda = 4$ per hour, $\mu = 60/12 = 5$ per hour

therefore the average time a machine spends in the system

$$= \frac{1}{\mu - \lambda} = \frac{1}{5 - 1} = 1 \text{ hour}$$

So the average lost production cost per machine $= 1 \times £50 = £50$. *But*, on

average, four machines break down per hour so the average lost
production per hour $= 4 \times £50$ $= £200$
Also, hourly cost of employing eight mechanics $= 8 \times £4$ $= £32$
                                                                        ———
                                    therefore the average total cost per hour $= £232$

For the nine-mechanic team:
$\lambda = 4$ per hour, $\mu = 60/10 = 6$ per hour
Therefore the average time a machine spends in the system

$$= \frac{1}{\mu - \lambda} = \frac{1}{6-4} = 0.5 \text{ hours}$$

So the average lost production cost per machine $= 0.5 \times £50 = £25$.
So the average lost production cost per hour $= 4 \times £25$ $= £100$.
The hourly cost of employing nine mechanics $= 9 \times £4$ $= £36$
                                                                       ———
Therefore the average total cost per hour $= £136$
Therefore it is worth employing a ninth mechanic.

11.4 Simulation

This topic appears only on the CACA(2.6) and CIPFA syllabuses

Simulation models are designed to mimic the behaviour of a real system
over time. For example, a model might be used to simulate the queues of
customers at supermarket checkouts over an eight-hour day.
   The main advantages of simulation are:
1   It can be used to model problems where a mathematical model would
    be extremely complex or impractical.
2   The decision maker can experiment with the simulation model to
    determine the best course of action (e.g. the effect of operating
    different numbers of supermarket checkouts could be observed).
    Experimenting with the real system may be expensive, dangerous or
    impossible.
3   It allows the behaviour of the system over a long period (e.g. a year) to
    be observed in a computer run taking only a few minutes.
   The main limitations of simulation are:
1   It is effectively sampling the behaviour of the system therefore a
    simulation result is only an estimate which is subject to sampling error.
    However, this error can be reduced by running the simulation for
    longer periods.
2   It normally requires the use of a computer.
3   Like any model, a simulation model will, to some extent, be a
    simplification of the real system.
4   It does not prescribe an optimal solution to a problem. The 'best'
    course of action is found by experimentation.

11.4.1 The stages in simulation

The application of simulation to a problem will probably involve the following steps:

1   Definition of the problem. This will require the full involvement of management so that the objectives of the simulation study can be set out.
2   Analysis of the real system. This will involve identification of the interactions between the different elements of the system and collection of data so that, for example, probability distributions can be defined.
3   The formulation of a computer model to represent the real system.
4   Testing of the model. Does it adequately represent the real system?
5   Experimentation so that the effects of different policies can be compared.
6   Implementation i.e. putting the results of the simulation study into action in the real system.

11.4.2 The use of random numbers in simulation

Random numbers (normally generated by a computer, or from a set of random number tables) can be used to simulate random behaviour in a system.

Example

The weekly demand for a liquid chemical product which is stocked by a company follows the probability distribution below. Simulate the demand for six weeks.

Demand per week (gallons)	Probability
0 to under 10	0.3
10 to under 20	0.6
20 to under 30	0.1
	1.0

1   The distribution can be simplified by replacing the demand classes by their midpoints (as shown below).
2   The probabilities are then cumulated.
3   If single digit random numbers in the range 0 to 9 are to be used, the first 3 numbers (i.e. 0, 1 and 2) will be allocated to a demand of 5 units. (Thus 3 of the 10 possible random numbers represent this level of demand which reflects the 3/10 probability of this level of demand occurring.) The next random numbers (3 to 8) are allocated to a demand of 15, and only one random number (9) represents a demand of 25.

Random numbers can be allocated as follows: (i) ignore the decimal points in front of the cumulative probabilities (ii) for the first class, the range of random numbers will be 0 to the cumulative probability less one (iii) for the next class, the range starts with the next available random number and finishes with the cumulative probability less one, and so on.

Weekly demand	Class midpoint	Probability	Cumulative probability	Random numbers
0 to under 10	5	0.3	0.3	0–2
10 to under 20	15	0.6	0.9	3–8
20 to under 30	25	0.1	1.0	9

4   If the following random numbers are generated:

    6   2   9   1   0   6

The simulated demand for the first 6 weeks would be:

    15   5   25   5   5   15 units

If the simulation run was for a much larger number of weeks, it could be used to compare the costs of, for example, operating with different re-order levels.

12
Network analysis

The material in this chapter is relevant to the following syllabuses: CIMA, ICAEW, ICAS, CACA(2.6), CIPFA

12.1 Introduction

A project is a set of activities which must be carried out in a defined order (e.g. building a garage will consist of the activities like laying the foundations, erecting the roof etc.). Network analysis can be used in the planning and control of projects, ensuring that the resources associated with the project (e.g. labour, machines and cash) are used effectively. Critical path analysis and PERT (project evaluation and review technique) are two of the main methods.

Note that sometimes the terms critical path analysis and PERT are used synonymously.

12.2 Networks and activity times

In critical path analysis a project is represented as a network diagram. In the 'activity-on-the-node' system each activity is represented by a node while arrows show the sequence in which activities will be carried out. Generally, the following information about an activity can be displayed on its node (e.g. Figure 12.1).

1 The earliest start time (EST) i.e. the earliest possible time at which an activity can start.
2 The earliest finish time (EFT) i.e the earliest possible time that an activity can be completed by. Clearly, it is equal to the EST plus the activity's duration.
3 The latest start time (LST) i.e. the latest time at which an activity can start without delaying the project's completion.

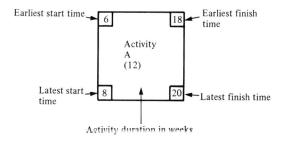

Figure 12.1

4   The latest finish time (LFT) i.e. the latest time at which an activity can finish without delaying the project's completion.

Suppose that the overhaul of a large machine consists of seven activities details of which are given in Table 12.1.

Table 12.1

Activity	Description	Estimated duration (days)	Preceding activities	No. of engineers required per day
A	Remove machine casing	1	none	3
B	Remove pumps	2	A	1
C	Clean cylinders	7	A	2
D	Replace filters	3	A	4
E	Clean and replace pumps	4	B	2
F	Replace casing	1	C, D and E	3
G	Test and adjust controls	1	F	2

An activity-on-the-node network for the project is shown in Figure 12.2. The calculation of the start and finish times is carried out as follows.

1   Calculate the ESTs and EFTs (i.e. the numbers at the tops of the nodes) moving from left to right through the network. Thus A can start at time zero and therefore it will finish after one day. This means that B, C and D can start after one day has elapsed, and so on. The key point to note is that if an activity is preceded by more than one activity, the start of that activity will be held up by the preceding activity which finishes last. Therefore the *highest* preceding EFT applies (e.g. F is preceded by E which finishes after seven days, C finishing after eight days and D finishing after four days. Thus F's earliest start time is eight days).

The EFT of the last activity, G, gives the duration of the project which is ten days.

2   Next calculate the LFTs and LSTs (i.e. the bottom numbers). Start with the last activity and work backwards throughout the network. Thus G must have a latest finish time of ten days to avoid delaying the project's completion. It must therefore start at the latest by day nine. This means that F must have a latest finish time of nine days, and so on.

The key point here is that, if an activity is followed by more than one activity, the *smallest* LST of the succeeding activities is selected. Thus A is followed by B (which must start at the latest by day two), C (LST = one day) and D (LST = five days). To avoid delaying the start of any of these activities A must be completed by the end of day one.

To illustrate this important point, consider a meeting where one person arrives at 1 p.m., another at 1.05 p.m. and a third at 1.35 p.m. If the meeting cannot start until all three people have arrived then its earliest start time is 1.35 p.m. i.e. the arrival time of the last person.

Note that the ESTs etc. are 'elapsed times' (i.e. the amount of time which has elapsed 'on the clock' since the start of the project). Thus A finishes after one day has elapsed and B starts after one day has elapsed. Thus B does not start during day one.

12.3 Float

The total float (sometimes simply called the float) of an activity is the amount of time by which its start (or completion) can be delayed without holding up the project. Therefore: total float = latest start time − earliest

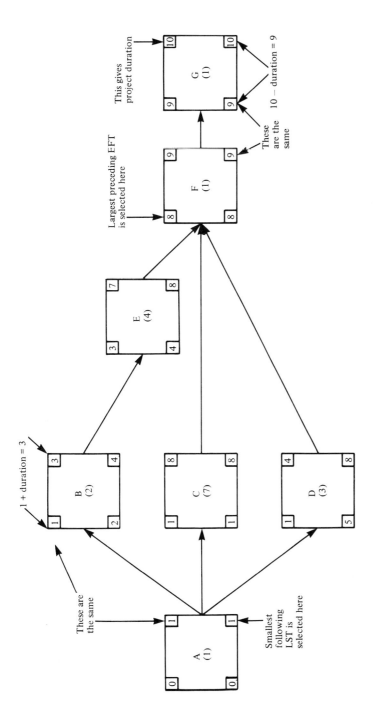

Figure 12.2

start time. The total floats of the activities in the overhaul project are

	A	B	C	D	E	F	G
Total float (days)	0	1	0	4	1	0	0

A, C, F and G have zero total float. Any delay in their start or completion will delay the entire project. They are therefore called *critical activities* and the path they lie on, A–C–F–G (the path with the largest sum of durations) is called the *critical path*.

12.4 Using critical path analysis for resource allocation

Critical path analysis can be used to make efficient use of the resources (in this case engineers) which are required to carry out a project. First a time chart is drawn to show when the activities will take place. In the chart in Figure 12.3 it has been assumed that all the activities start at their earliest possible time. The Xs show that some of the activities can, if necessary, be 'pushed to the right' (i.e. delayed) by a certain number of days. The number of engineers required per day (as shown on the bar chart) is calculated by summing the requirements of the activities which take place

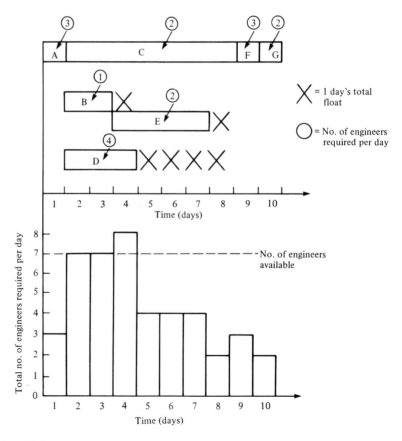

Figure 12.3

on each day. For example, on day four, C (requiring two engineers), E (two engineers) and D (four engineers) take place, giving a total requirement of eight engineers.

If only seven engineers are available, can the project still be completed in ten days? The answer is yes because the start of E can be delayed until the end of day four, and a maximum of seven engineers would then be required on any one day.

12.5 Speeding up (or crashing) activities

Sometimes it is necessary to consider speeding up the completion of activities (e.g. in order to meet a deadline). However, the extra costs incurred through speeding up activities (e.g. the payment of overtime to employees) need to be balanced against the costs of late completion (e.g. penalty costs in a contract).

Suppose that if the machine overhaul is not completed within seven days, production will be lost at a cost of £1,000 per day. The crash (i.e. minimum possible) durations of activities are given below. The table also shows what it will cost to reduce the completion time of an activity by a day. Is it worth speeding up any of the activities?

Activity	*A*	*B*	*C*	*D*	*E*	*F*	*G*
Normal duration (days) (as before)	1	2	7	3	4	1	1
Crash duration (days)	1	1	4	1	2	1	1
Cost per day saved (£)	–	£100	£700	£200	£500	–	–

The three 'golden rules' which apply here are
1   It is only worth speeding up critical activities (the other activities are not holding the project up).
2   If there is more than one critical path, activity durations on each path must be reduced together (otherwise one of the paths will still be at its original length and the overall project duration will not have been reduced).
3   It is not worth speeding up an activity if this will cost more than the saving which will be made by finishing the project earlier. Also, when there is a choice of activities to speed up, always select the cheapest alternative.

These rules can be applied to the machine overhaul project as follows.

Step 1
The critical activities are A, C, F and G. Only C's duration can be reduced, at a cost of £700 per day saved. However, since speeding up C by one day will save £1,000, it is worth carrying out. Figure 12.4 shows the result. Note that the project duration is now nine days. However, there are now two critical paths since A–B–E–F–G also takes nine days.

For simplicity the ESTs etc. have not been calculated here. The critical path has simply been identified as the path with the greatest total duration.

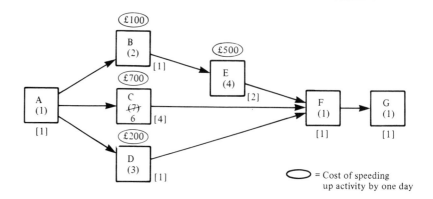

£100

B
(2)
[1]

£700

C
(7̶)
6
[4]

£500

E
(4)
[2]

A
(1)
[1]

£200

D
(3)
[1]

F
(1)
[1]

G
(1)
[1]

⬭ = Cost of speeding
up activity by one day

[ ] = Crash duration in days

Figure 12.4

Step 2

C's duration can be reduced further. However, either B or E must be reduced at the same time

reducing C by 1 day and B by 1 day costs £700 + £100 =   £800
reducing C by 1 day and E by 1 day costs £700 + £500 = £1,200

Therefore reduce C and B by one day to give a project duration of eight days.

Step 3

C's duration can still be reduced further, but B is at its crash duration. The only option is to reduce C and E by one day, but (as shown in step 2) this will cost £1,200 which is greater than the £1,000 saving which will be made. Therefore no further reductions are worth considering. The optimal project duration is eight days.

12.6 PERT

In critical path analysis the activity durations are generally only estimates and unforeseen circumstances can often reduce or extend the completion time of activities. The PERT method uses probability theory to take into account this uncertainty about activity durations.

Three estimates of activity durations are made: an optimistic estimate (OPT), a 'most likely' estimate (M) and a pessimistic estimate (PESS). This has been done for the activities of the network shown in Figure 12.5. For each activity an expected duration is then calculated as a weighted average of these three estimates (M has a weighting of 4, the other estimates a weighting of 1). Thus

$$\text{Expected duration} = \frac{\text{OPT} + (4 \times \text{M}) + \text{PESS}}{6}$$

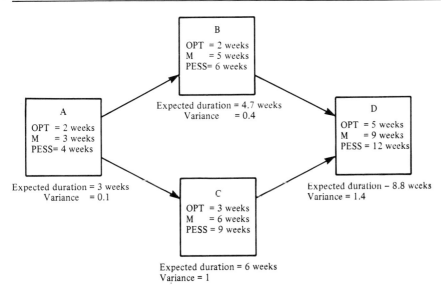

Figure 12.5

Then, for each activity, the variance of the estimates is calculated using the following formula

Note that a variance of zero implies that we are certain about an activity's duration.

$$\text{Variance of activity duration} = \left(\frac{\text{PESS} - \text{OPT}}{6}\right)^2$$

For the activities in Figure 12.5 the results obtained are

Activity	A	B	C	D
Expected duration (weeks)	$\frac{2+4\times3+4}{6} = 3$	$\frac{2+4\times5+6}{6} = 4.7$	$\frac{3+4\times6+9}{6} = 6$	$\frac{5+4\times9+12}{6} = 8.8$
Variance	$\left(\frac{4-2}{6}\right)^2 = 0.1$	$\left(\frac{6-2}{6}\right)^2 = 0.4$	$\left(\frac{9-3}{6}\right)^2 = 1$	$\left(\frac{12-5}{6}\right)^2 = 1.4$

Next, the following assumptions are made

1   The path with the greatest sum of expected durations is the critical path

If more than one path fits this criterion then the path with the largest variance is assumed to be the critical path.

      A–B–D = 3 + 4.7 + 8.8 = 16.5 weeks
      A–C–D = 3 + 6 + 8.8   = 17.8 weeks
      so A–C–D is the critical path

2   The activity durations are independent of each other (e.g. if one activity takes longer than expected it will not affect the probability of other activities being completed within a certain time). This means that the variances of the activity durations can be added.

Thus the variance of the time taken to complete the critical path = variance of A + variance of C + variance of D = 0.1 + 1 + 1.4 = 2.5. Therefore the standard deviation = $\sqrt{2.5}$ = 1.6 weeks (approximately).

3   The overall project duration will be normally distributed (this may not be the case where there are only a few activities on the critical path). Thus the project completion time is normally distributed with a mean (expected) duration of 17.8 weeks and a standard deviation of 1.6 weeks.

This means that the probability of the project taking certain periods of time to complete can be calculated. For example, to find the probability that the project will take longer than twenty weeks to complete

$$Z = \frac{20 - 17.8}{1.6} = 1.38$$

so the appropriate probability (from the normal tables) is 0.0838.

Examination questions

Answers to questions which involve calculations are given on page 138.

1 Statistical sampling techniques are widely used for the collection of data in industry and business. Explain four of the following, illustrating your answer with examples

(a) sampling frame;
(b) simple random sampling;
(c) multi-stage sampling;
(d) stratification;
(e) quota sampling;
(f) sampling with probability proportional to size.

(CIMA, specimen paper for new syllabus 1986)

2 (Multiple choice question)

Two equations are given by

$$3Y = 2 + X$$
$$2Y = 28 - 2X$$

These lines cut at the following coordinates

A: $X = -10, Y = 4$
B: $X = -2, Y = 0$
C: $X = 0, Y = 2/3$
D: $X = 10, Y = 4$
E: None of the above
(Select the letter which represents the answer of your choice.)

(CIMA, Mathematics and Statistics, May 1985)

3 The finance division of a large company is investigating its procedures for the selection of new accountancy trainees. Potential applicants are given, prior to appointment, both a written test and a formal interview. The performances of eight successful applicants were rated after their first full year with the company. The independent rankings of written test, interview assessment and job performance for the eight trainees are given below

(1 = best, 2 = second best, etc.)

Trainee	A	B	C	D	E	F	G	H
Written test	6	2	7	4	1	5	3	8
Interview	1	4	2	3	6	5	8	7
Job performance	1	2	3	4	5	6	7	8

$$\text{Spearman's } r' = 1 - \frac{6\Sigma d^2}{n(n^2 - 1)}$$

You are required to

(a)   Calculate a rank correlation coefficient between
    (i) job performance and written test;
    (ii) job performance and interview assessment;
(b)   interpret your results.

(CIMA, *Mathematics and Statistics*, May 1985)

4 (a)   Your company requires a special type of inelastic rope which is available from only two suppliers. Supplier A's ropes have a mean breaking strength of 1,000 kg with a standard deviation of 100 kg. Supplier B's ropes have a mean breaking strength of 900 kg with a standard deviation of 50 kg. The distribution of the breaking strengths for each type of rope is normal. Your company requires that the breaking strength of a rope be not less than 750 kg.

All other things being equal, which rope should you buy, and why?

(b)   One per cent of calculators produced by a company is known to be defective. If a random sample of fifty calculators is selected for inspection, calculate the probability of getting no defectives by using: (i) the Binomial distribution; (ii) the Poisson distribution.

(CIMA, *Mathematics and Statistics*, May 1984)

5   A company has the opportunity of marketing a new package of computer games. It has two possible courses of action: to test market on a limited scale or to give up the project completely. A test market would cost £160,000 and current evidence suggests that consumer reaction is equally likely to be 'positive' or 'negative'. If the reaction to the test marketing were to be 'positive' the company could either market the computer games nationally or still give up the project completely. Research suggests that a national launch might result in the following sales:

Sales	Contribution £ million	Probability
High	1.2	0.25
Average	0.3	0.5
Low	−0.24	0.25

If the test marketing were to yield 'negative' results the company would give up the project. Giving up the project at any point would result in a contribution of £60,000 from the sale of copyright etc. to another manufacturer. All contributions have been discounted to present values.

You are required to
(a) draw a decision tree to represent this situation, including all relevant probabilities and financial values;
(b) recommend a course of action for the company on the basis of expected values;
(c) explain any limitations of this method of analysis.

(CIMA, specimen paper for new syllabus 1986)

6 A client company of your firm is a horticultural shop selling a wide variety of products to its customers. The analysis of weekly sales of plants throughout the year is summarized in the following frequency distribution.

Weekly sales of plants £	Number of weeks
1,255 but less than 1,280	9
1,280 but less than 1,305	19
1,305 but less than 1,330	10
1,330 but less than 1,355	8
1,355 but less than 1,380	6

Required
(a) Construct a fully labelled histogram of the frequency distribution.
(b) Establish (to the nearest £) the value of the mode from the histogram using a graphical method.
(c) Calculate the mean weekly sales of plants over the fifty-two weeks to the nearest £.
(d) Compare and contrast the value of the mode and mean of this frequency distribution to the manager of the horticultural shop.
(e) Calculate the standard deviation of the frequency distribution.

(CACA, Paper 1.5 *Numerical Analysis and Data Processing*, December 1985)

7 Prodco PLC manufactures an item of domestic equipment which requires a number of components which have varied as various modifications of the model have been produced.

The following table shows the number of components required together with their price over the last three years of production.

Component	1981		1982		1983	
	Price	Quantity	Price	Quantity	Price	Quantity
	£		£		£	
A	3.63	3	4.00	2	4.49	2
B	2.11	4	3.10	5	3.26	6
C	10.03	1	10.36	1	12.05	1
D	4.01	7	5.23	6	5.21	5

Required

(a) Establish the base weighted price indices for 1982 and 1983 based on 1981 for the item of equipment.

(b) Establish the current weighted price indices for 1982 and 1983 based on 1981 for the item of equipment.

(c) Using the results of (a) and (b) as illustrations compare and contrast Laspeyres and Paasche price index numbers.

(CACA, Paper 1.5 *Numerical Analysis and Data Processing*, December 1984)

8   The group chief accountant of the Nisus Car Company has asked you to arrange a training course on 'Financial Planning Using Microcomputers' for the accounts staff within the group's six subsidiary companies. Having decided upon a two-week course, you list the various jobs which will need doing, along with their estimated times and any necessary preceding jobs as follows:

	Job	Time (weeks)	Preceding jobs
A	Arrange programme	2	–
B	Prepare publicity	3	A
C	Publicize course	2	B
D	Receive applications	4	C
E	Prepare course material	8	A
F	Decide on location of course	1	D
G	Arrange accommodation	2	F
H	Arrange computer facilities	3	F
I	Run course	2	E, G, H
J	Set course assignments	3	D, E
K	Mark assignments	2	I, J
L	Post-course evaluation	1	I, K

Required

(a) Assuming that the work of organizing the course can start immediately, what is the minimum number of weeks that must elapse before the course can be run?

(b) After the post-course evaluation, you must report back to the group chief accountant on the success of the course. How long will it take (i.e. how many weeks) before you can begin to compile your report?

(c) Comment on how uncertainty might be analysed within a situation such as this.

(CACA, Paper 2.6 *Quantitative Analysis*, December 1984)

9   As a result of a routine analysis of cash flows, the chief accountant of Odin Chemicals Limited considers that there are only three types of cash flow which are likely to vary significantly from month to month. These are wages and salaries, raw materials purchases, sales revenue.

Using data that has been collected over the last two years, and taking into account the likely changes in the level of operations during the next few months, the following distributions have been estimated for the monthly cash flow in each of these three categories.

Wages and salaries	probability	Raw materials	probability
(£000)		(£000)	
10–12	0.3	6–8	0.2
12–14	0.5	8–10	0.3
14–16	0.2	10–12	0.3
		12–14	0.2

Sales revenue	Probability
(£000)	
30–34	0.1
34–38	0.3
38–42	0.4
42–46	0.2

All the other cash flows can be regarded as fixed, and amount to a net outflow of £14,000 per month. Currently Odin has cash assets of £50,000.

Required

(a) Using the random numbers given at the end of the question, simulate six months' cash flows. You may assume that all cash flows are independent and take place at the end of the month. From your simulation, estimate the probability of a net cash outflow in any month, and the cash balance at the end of the six-month period.

(b) What is the *expected* cash balance at the end of the six-month period? Explain briefly why this value may be different from a corresponding value obtained from a simulation experiment.

(c) If a more comprehensive and realistic model is to be constructed, discuss carefully and concisely the points which must be considered when simulating cash flows over a period of time.

Random numbers: Wages and salaries 2 7 9 2 9 8
Raw materials 4 4 1 0 3 4
Sales revenue 0 6 6 8 0 2

(CACA, Paper 2.6 *Quantitative Analysis*, December 1983)

10 The following figures relate to the sales per calendar quarter of a trading company over the four years 1981–4

Year	Quarters (£000s)			
	I	II	III	IV
1981	78	84	94	84
1982	87	92	97	91
1983	93	99	106	101
1984	104	109	117	112

You are required to
(a)   estimate the seasonal movement;
(b)   estimate the trend of turnover for each of the four years;
(c)   derive estimates of the probable quarterly sales in 1985; and
(d)   outline the basis underlying this method of time series analysis, indicating its limitations.

(ICAEW, *Quantitative Techniques*, November 1985)

11   (a)   You are required to
(i) define and explain the term 'confidence interval'; and
(ii) calculate a 95 per cent confidence interval where 52 per cent of a sample of 800 customers state a preference for a given product against its competitors. Interpret your results.

(b)   In a controlled experiment the success rate of a new drug in curing a sample of 200 patients suffering from a certain disease is 60 per cent. The drug currently in use has a comparable success rate of 54 per cent. Are the doctors justified in inferring that the new drug is significantly superior as a treatment? Explain carefully how you interpret the result of your statistical test.

(ICAEW, *Quantitative Techniques*, November 1985)

12   The annual turnover and net profits of a company over the past decade are as follows

Year:		1	2	3	4	5	6	7	8	9	10
Turnover:	£m	60	75	90	120	140	145	150	160	170	140
Profit:	£m	3	5	7	10	13	16	12	15	18	20

You are required to
(a)   calculate the regression equation of profit on turnover;
(b)   estimate from the equation the profit expected in the next two years if turnover rises by 10 per cent in each year; and
(c)   comment on the reliability of your results in (b).

(ICAEW, *Quantitative Techniques*, May 1985)

13   A manufacturer can produce two products, A and B, requiring the following inputs

	A	B
Materials in kilos	2.25	1.80
Manufacturing time in hours	7	10
Finishing and packaging time in hours	10	6

The materials cost £4 per kilo and there are 180 kilos available. Manufacturing time costs £3 per hour and there are 840 hours available.

Finishing and packing time costs £2.50 per hour and there are 720 hours available. The selling price of both A and B is £70.

You are required to

(a) express the constraints if the objective is to maximize profits;

(b) use graphical methods to determine the output which would maximize profits; and

(c) advise the manufacturer as to any limitations of your proposed solution and make any recommendations which may enable profits to be increased.

(ICAEW, Quantitative Techniques, November 1984)

14 (a) Explain, in language comprehensible to a layman, the following terms and the application and advantages of each in practice (i) net present value (ii) internal rate of return.

    (b) Your company purchases a machine costing £50,000. Half the purchase price is payable on delivery and the balance a year later. The cash flow from its operation is expected to be £15,000 per annum. At the end of five years the machine will be sold – the expected price is £5,000.

        If the company has to borrow at 15 per cent per annum to finance the purchase of this machine, do you consider the purchase to be justified on financial grounds?

(ICAEW, Quantitative Techniques, November 1984)

Answers to examination questions

2 D is the correct answer.

3 (a) (i) 0.119 (ii) 0.881.

4 p(supplier A's rope has breaking strength > 750 kg) = 0.9938
P(supplier B's rope has breaking strength > 750 kg) = 0.9987
therefore: buy B's ropes.
(i) 0.605 (ii) 0.6065

5 Optimal sequence of decisions is: test market and, if the results are 'positive', market nationally (highest EMV = £65,000).

6 (b) £1,293 (c) £1,309 (e) £31.

7 (a) 1982: 124.3, 1983: 130.6 (b) 1982: 125.7, 1983: 133.3.

8 (a) after fifteen weeks (b) after twenty weeks.

9 (a) p(net cash outflow) = 0.5, £60,000 (b) £62,000.

10 (a) (b) Centred averages (to 1 decimal place):

	1981				1982			
Qtrs	1	2	3	4	1	2	3	4
			86.1	88.5	89.6	90.9	92.5	94.1

	1983				1984			
Qtrs	1	2	3	4	1	2	3	4
	96.1	98.5	101.1	103.8	106.4	109.1		

Seasonal deviations (from additive model):
Q1: -2.7, Q2: 0.5, Q3: 5.7, Q4: -3.5 (to 1 decimal place).
(c) 1985 forecasts (£000s) Q1: 112, Q2: 118, Q3: 124, Q4: 118 (these may vary since much depends on the judgement of the forecaster).

11 (a) (ii) 52% ± 3.46%
(b) $H_0 \pi = 54\%$ $H_1 \pi > 54\%$; $Z = 1.70$ so reject H_0 at 5% level of significance, but not at 1%.

12 (a) $y = -4.93 + 0.135x$ (b) Year 11: £15.86m; Year 12: £17.94m (results may differ slightly, depending on the degree of rounding).

13 The model is:

Maximize profit $= 15a + 17.8b$
subject to $2.25a + 1.8b \le 180$
$7a + 10b \le 840$
$10a + 6b \le 720 \; a, b \ge 0$

(where $a =$ number of A produced and $b =$ number of B produced).
To maximize profit: $a = 29.1, b - 63.6$.

14 (b) NPV $= +£6,030$ therefore the purchase does seem justifiable.

Part Three
Exercises and Answers

Exercises

Worked answers to these questions are given on pages 150 to 163.

Chapter 1

Question 1.1

A food company supplies a product to restaurants, shops and wholesalers in the London area. The company recently changed its delivery system and it intends to carry out a survey of customers to find out if they prefer the new system. The survey will involve a company representative visiting and interviewing a sample of customers. How should the company select this sample?

Question 1.2

CACA(2.6) students do not attempt this question

The value of goods purchased by thirty customers at a department store is given below.
(a) Organize the data into a frequency distribution.
(b) Draw a histogram and ogive to represent the distribution.

£29.50	£6.00	£4.98	£79.00	£38.00	£37.00
£56.19	£21.06	£43.10	£15.20	£39.99	£65.00
£26.14	£53.21	£32.00	£39.00	£50.20	£73.00
£59.00	£38.00	£45.00	£36.40	£16.00	£96.00
£28.00	£29.00	£21.20	£50.00	£53.20	£33.40

Chapter 2

Question 2.1

A company is involved in wage negotiations with the trades unions. The current gross weekly wages of the company's employees are summarized in the frequency table below. Two proposals are on the table:

Proposal 1: Give the employees an increase in pay equivalent to 10 per cent of the current mean weekly wage.

Proposal 2: Give the employees an increase in pay equivalent to 12.5 per cent of the current median weekly wage.

Compare the costs of the two proposals.

Gross weekly wages (£s)	No. of employees
90 to under 110	14
110 to under 140	54
140 to under 160	16
160 to under 200	12
200 and over	2
	98

Question 2.2

The frequency distribution below shows the number of employees absent from work at an insurance company office over a thirty-day period during this year.

Number of employees absent	No. of days
0	4
1	22
2	3
3	1
	30

For the equivalent thirty-day period last year the mean number of employees absent was 2.9 with a standard deviation of 1.9. Compare this year's figures with last year's.

Chapter 3

Question 3.1

CACA(1.5) students should not answer this question

The total expenses (in £000s), paid to a company's sales representatives for each quarter of 1984, 1985 and 1986, are shown below.

Year	1984				1985				1986			
Quarter	1	2	3	4	1	2	3	4	1	2	3	4
Expenses	5	9.2	10.8	3	6.5	9	12.1	4	7	10	14	6

(a) Use the method of moving averages to find the trend in the data.
(b) Use the additive time series model to find the seasonal deviations and interpret your results.

Question 3.2

ICAS, CACA(2.6) and CIPFA students should not attempt this question

Each year since 1982 a Laspeyre index number has been calculated to monitor a company's production costs. The indices for the years 1982–5 (with 1982 as base year) are given below.

Year	1982	1983	1984	1985
Cost index	100	102	109	116

The production costs consist mainly of labour, raw material and fuel costs and details of these costs for 1982 and 1986 are given below together with the number of units of each resource which were used in each year.

	1982		1986	
	Cost per unit	No. of units used	Cost per unit	No. of units used
Labour	£1.90 per hour	10,400 hours	£2.42 per hour	9,420 hours
Raw materials	£18 per tonne	500 tonnes	£22 per tonne	520 tonnes
Fuel	£2 per litre	5,000 litres	£2.40 per litre	4,900 litres

(a) Calculate the cost index for 1986, using 1982 as base year.
(b) A price index for the whole economy is given below for the years 1982–6 (the base year for this index is 1984). Compare the rise in the company's production costs with the general rise in prices over this period.

Year	1982	1983	1984	1985	1986
Price index	86	90	100	112	114

Chapter 4

Question 4.1

A construction project involves two main activities, A and B, which are to be started simultaneously. The completion times of the activities follow the probability distributions shown below.

Activity A		Activity B	
Completion time (weeks)	Probability	Completion time (weeks)	Probability
4	0.1	3	0.2
5	0.7	4	0.2
6	0.2	5	0.6

(a)   Assuming that the completion times of the two activities are independent, find the probability that
  (i) both activities will take exactly five weeks to complete;
  (ii) both activities will be completed in four weeks or less;
  (iii) the two activities will have the same completion time;
  (iv) activity A takes longer than activity B.
(b)   Find the expected completion time of each activity and interpret your results.

Question 4.2

ICAI, CACA(1.5) and ICSA students should not attempt this question

A person who has a large sum of money to invest for the next twelve months has to choose between three types of investment, A, B or C. Investment A will definitely give returns of £5,000 over the next year, but the returns on investments B and C depend upon changes in interest rates. If interest rates fall, B will give a return of £3,000 and C a return of £1,000. However, if interest rates remain the same or rise, B will give a return of £7,000 and C a return of £8,000. It is estimated that there is a 'a 40 per cent chance' that rates will fall. Which investment will give the maximum expected payoff?

Chapter 5

Question 5.1

A company which delivers a good to customers using its own vehicle fleet agrees to replace, at a cost of £40, any unit which is damaged in transit. Experience suggests that approximately 2 per cent of units are damaged during delivery. In a given week the company is due to deliver 120 units.
(a)   Assuming that the Poisson approximation to the binomial distribution is valid, what is the probability that replacement costs resulting from the week's deliveries will exceed £100?
(b)   If the Poisson approximation to the binomial distribution is valid here, what does this imply?

Chapter 6

Question 6.1

A quality control inspector takes a simple random sample of thirty bags of cement from a warehouse. These bags have a mean weight of 10 kg with a standard deviation of 2 kg.
(a)   Estimate the mean weight of all the bags in the warehouse at the 95 per cent level of confidence and interpret your result.
(b)   If the inspector needs to estimate the mean weight of the bags to within ±0.5 kg at the 95 per cent level of confidence, how large a sample will he need to take?

Question 6.2

CIMA students should not attempt this question

A long-distance bus company claimed that only 12 per cent of their buses arrived at their destination more than five minutes late during August 1986. However, 200 randomly selected bus journeys were monitored during this period by an independent organization and thirty of the buses arrived more than five minutes late. Does this suggest that the bus company's claim is exaggerated?

Question 6.3

Only CACA(2.6), CIPFA and ICAI students should attempt this question

A company supplies goods to customers in Northern England, Scotland and the Midlands. One hundred randomly selected customers were asked if they had experienced any problems with late delivery of goods in the last month. The results are given below. Is there any evidence that the proportion of all customers experiencing late delivery varies between the three sales regions?

	Sales region			
	Northern England	Scotland	Midlands	Total
Not experienced late delivery	12	30	12	54
Experienced late delivery	8	20	18	46
Total	20	50	30	100

Chapter 7

Question 7.1

The total cost per month (C) of manufacturing a product is given by the equation

$$C = 300 + 24x$$

where x = the number of units produced per month.
   The total revenue per month (R) is given by the equation

$$R = -2x^2 + 80x$$

(a)   Derive an equation which shows how profit per month is related to the number of units produced.
(b)   Find the level of output per month which will maximize profits.
(c)   Find the level(s) of output per month at which the company will break even.

Chapter 8

Question 8.1

Only CIMA and CACA(1.5) students should attempt this question

(a) If $A = \begin{bmatrix} 20 & 4 \\ 2 & 18 \end{bmatrix}$ and $B = \begin{bmatrix} 5 & 3 \\ 4 & 0 \end{bmatrix}$

Find (i) $A + B$ (ii) $A \times B$.

Question 8.2

A company is considering the purchase of a road vehicle which will cost £30,000. The vehicle's operators expect that the inflow of cash resulting from the vehicle's operations would be £25,000 for four years while the outflow would be £16,000 per year. At the end of year 4 the vehicle would be sold for an estimated price of £6,000. If the cost of capital is 12 per cent, should the vehicle be purchased?

Question 8.3

Money loaned to a local authority earns 8 per cent per annum compound interest for the first three years of the loan and 10 per cent thereafter. If £2,000 is loaned to the authority now, how much will it have to repay when the loan is terminated after five years?

Chapter 9

Question 9.1

The mean weekly temperature at midday (as recorded by a local weather centre) and the weekly heating costs of an office block are shown below for eight randomly selected weeks.

Mean temperature (degrees C)	(x)	10	5	0	11	15	17	25	20
Heating costs (£00s)	(y)	4	6.9	8	5	3.1	1	0	0.9

(a) Plot the results on a scattergraph.
(b) Given that $\Sigma x = 103$, $\Sigma y = 28.9$, $\Sigma xy = 211$, $\Sigma x^2 = 1,785$ and $\Sigma y^2 = 164.03$ calculate Pearson's product–moment coefficient of correlation for the data and interpret your result.
(c) Calculate the coefficient of determination for the data and interpret your result.
(d) Calculate the least-squares regression line and superimpose it on your scattergraph. Explain what the line shows about the relationship between temperature and heating costs.

(e) What would be the expected heating costs for a week when the mean midday temperature was 16 degrees centigrade?

Question 9.2

Seven accounting packages are ranked by a magazine according to how user-friendly they are (1 = most user-friendly, 7 = least user-friendly). These rankings are given below, together with the cost of the packages. Is there any evidence that user-friendliness is associated with cost?

Package	A	B	C	D	E	F	G
User-friendliness rank	5	2	1	4	3	6	7
Cost	£2,600	£2,400	£4,100	£3,500	£3,000	£900	£1,000

Chapter 10

Question 10.1

A day's feed for a farm animal must contain at least eight units of protein, twelve units of iron and nine units of vitamin D. The feed can be made up of two ingredients, *X* and *Y*. A kilogram of ingredient *X* contains two units of protein, six units of iron and one unit of vitamin D. A kilogram of ingredient *Y* contains one unit of protein, one unit of iron and three units of vitamin D.

Ingredient *X* costs £1.70 per kilogram while ingredient *Y* costs £0.80 per kilogram. Determine the number of kilograms of the two ingredients which must be used per day to minimize the cost of feeding the animal.

Question 10.2

Only CIPFA and CACA(2.6) students should attempt this question

A company assembles three types of cooker: the Standard, the De-luxe and the Cordon-bleu. The number of cookers which can be assembled in a week is limited by the total assembly time available (228 person-hours per week) and the number of type E components which can be obtained (these are temporarily in short supply and only eighty are available per week). The assembly time of the cookers and the number of type E components required by each cooker are shown below.

	Model		
	Standard	De-luxe	Cordon-bleu
Assembly time (person-hours)	4	5	6
No. of type E components required	1	1	2

The contribution earned on a Standard cooker is £20 while the De-luxe and Cordon-bleu models earn £25 and £40 respectively. The company need to determine the number of cookers of each type which should be assembled in a week in order to maximize contribution.

(a)   Formulate a linear programming model to represent the problem.
(b)   A computer package, which is used to solve the model, produces the following simplex tableau.

x_1	x_2	x_3	S_1	S_2	
0.67	0.83	1	0.17	0	38
−0.33	−0.67	0	−0.33	1	4
6.67	8.33	0	6.67	0	1,520

where x_1 = the number of Standard models produced per week
x_2 = the number of De-luxe models produced per week
x_3 = the number of Cordon-bleu models produced per week
S_1 = the number of spare person-hours available per week
S_2 = the number of spare type E components available per week

(i)   How many units of each model should be produced per week?
(ii)   Are there any spare person-hours or type E components at the optimum solution?
(iii) What is the weekly contribution at the optimum solution?
(iv)   What are the shadow prices of person-hours and type E components?
(v)   By how much must the contribution per unit of the three types of cooker increase in order to make them worth assembling.

Chapter 11

Question 11.1

CACA(1.5) students should not attempt part (b) of this question

(a)   The weekly demand for a product, which is held in stock, is always nine units per week. When an order for new stock is placed, it is delivered almost instantaneously, but an ordering cost of £50 is incurred. The product costs £30 per unit, and it costs about 15 per cent of this amount to hold a unit in stock for a year. Currently orders are placed for 200 units. What saving (if any) could be made by changing this re-order quantity? (Assume that there are fifty weeks in a year.)
(b)   The weekly demand for a second product is approximately normally distributed with a mean of twenty units and a standard deviation of five units, and the demand in any one week is independent of all other weeks. The time between placing an order for new stock and receiving the goods is four weeks. If the objective is to run out of stock in no more than 3 per cent of lead times, what should the re-order level be?

Question 11.2

Only CIPFA and CACA(2.6) students should attempt this question

The number of lorries arriving at an unloading bay follows a Poisson distribution with a mean of four per hour. Currently, each lorry is unloaded by a team of three people who are each paid £2.50 per hour. On average it takes this team ten minutes to unload a lorry, but the actual time follows a negative exponential distribution. Each hour a lorry spends either queuing or being unloaded costs an estimated £6 per hour and the management are considering increasing the team to four people in order to reduce delays. It has been estimated that a team of four people could unload a lorry in an average time of eight minutes. Should the size of the team be increased?

(Assume that all the assumptions of the (M/M/1) model are valid here.)

Chapter 12

Question 12.1

A project consists of nine activities, details of which are given below:

Activity	Estimated duration (weeks)	Preceding activity
A	6	–
B	5	A
C	4	A
D	1	A
E	3	B and C
F	7	C and D
G	2	D
H	9	E and F
I	4	F and G

(a) Draw a network to represent the project.
(b) For each activity, calculate the earliest and latest start and finish times and the total float.
(c) Identify the critical path. What is the estimated duration of the project?
(d) What would be the effect of:
   (i) eight weeks elapsing before activity B can start
   (ii) activity F taking nine weeks instead of the estimated seven?

Answers to exercises

Question 1.1

Since a list of customers will probably be available, this can be used as the sampling frame. It is possible that the three types of customer will have reacted differently to the new delivery system so it is necessary to ensure that all three categories are represented in the sample. This could be achieved by taking a stratified sample. The customers are all located in the London area so the stratified sample would not require the representatives to travel large distances in order to carry out the interviews.

Question 1.2

The smallest value is £4.98 and the highest is £96. This is roughly a range of £0 to £100 which can be conveniently divided into five classes with a class interval of £20. (Of course, this is largely a matter of personal judgement and other answers would be valid.) A tally can then be used to count the number of values falling into each class.

Value of purchases (£)	Tally	Number of customers (= frequency)
0 to under 20	////	4
20 to under 40	ЦНŤ ЦНŤ ////	14
40 to under 60	ЦНŤ ///	8
60 to under 80	///	3
80 to under 100	/	1
	Total	30

The cumulative frequency distribution is

Value of purchases (£)	Cumulative frequency
Under 20	4
Under 40	18
Under 60	26
Under 80	29
Under 100	30

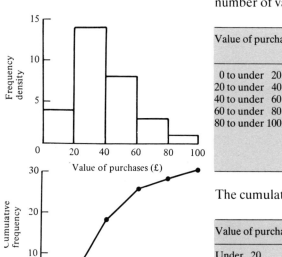

Figure A1

The histogram and ogive are shown in Figure A1.

Question 2.1

Assuming that the £200 and over class is as wide as the preceding class, it becomes £200 to under £240.

Proposal 1

Gross weekly wages (£s)	f	x = class mid-point	fx
90 to under 110	14	100	1,400
110 to under 140	54	125	6,750
140 to under 160	16	150	2,400
160 to under 200	12	180	2,160
200 to under 240	2	220	440
	$\Sigma f = 98$		$\Sigma fx = 13{,}150$

Therefore

$$\text{the mean} = \frac{\Sigma fx}{\Sigma f} = \frac{13{,}150}{98} = £134.18$$

so each employee would receive an increase of 10 per cent of £134.18 i.e. £13.42. Since there are ninety-eight employees, this would cost an extra £13.42 × 98 = £1,315 per week (i.e. 10 per cent of the total wage bill Σfx).

Proposal 2

The cumulative frequency distribution is

Gross weekly wages (£s)	No. of employees (= cumulative frequency)
Under 110	14
Under 140	68
Under 160	84
Under 200	96
Under 240	98

Forty-nine (i.e. 98/2) employees earn less than the median wage. An ogive would reveal that the median is approximately £129. This would imply a wage increase of 12.5 per cent of 129 = £16.13 (approx.) per worker. Paying this to ninety-eight employees would cost an extra £16.13 × 98 = £1,581 per week and would therefore cost more than proposal 1.

Question 2.2

The number of absent employees is a discrete variable and the frequency distribution has not been grouped into classes. Therefore, there is no need to find class mid-points when calculating the mean.

No. of employees absent ($=x$)	No. of days ($=f$)	fx	fx^2
0	4	0	0
1	22	22	22
2	3	6	12
3	1	3	9
	$\Sigma f = 30$	$\Sigma fx = 31$	$\Sigma fx^2 = 43$

$$\text{Mean number of employees absent} = \frac{\Sigma fx}{\Sigma f} = \frac{31}{30} = 1.03 \text{ employees}$$

$$\text{Standard deviation} = \sqrt{\frac{\Sigma fx^2}{\Sigma f} - \left(\frac{\Sigma fx}{\Sigma f}\right)^2} = \sqrt{\frac{43}{30} - \left(\frac{31}{30}\right)^2}$$

$$= 0.6 \text{ employees}$$

Therefore, the mean number of employees who were absent during the thirty-day period this year is lower than last year's figure. Also the variation in the number of employees absent per day is less this year. The relative variation can be measured by calculating the coefficient of variation.

$$\text{Last year's coefficient of variation} = \frac{\text{std dev.}}{\text{mean}} = \frac{1.9}{2.9} \times 100$$

$$= 65.5\%$$

$$\text{This year's coefficient of variation} = \frac{0.6}{1.03} \times 100 = 58.2\%$$

Question 3.1

(a) (c)	Year	Qtr	Expenses	4 quarterly moving average	Centred average	Seasonal deviation
	1984	1	5.0			
		2	9.2			
				7.000		
		3	10.8		7.19	3.61
				7.375		
		4	3.0		7.35	−4.35
				7.330		
	1985	1	6.5		7.49	−0.99
				7.150		
		2	9.0		7.78	1.22
				7.900		
		3	12.1		7.96	4.14
				8.030		
		4	4.0		8.15	−4.15
				8.280		
	1986	1	7.0		8.51	−1.51
				8.750		
		2	10.0		9.00	1.00
				9.250		
		3	14.0			
		4	6.0			

The trend in expenses paid is upwards, though, to some extent, this may reflect inflation.

Year	Q1	Q2	Q3	Q4
1984			3.61	−4.35
1985	−0.99	1.22	4.14	−4.15
1986	−1.51	1.00		

	Q1	Q2	Q3	Q4	Total	Adjustment
Mean	−1.25	1.11	3.88	−4.25	−0.51	+0.51/4
Adjustment	+0.13	+0.13	+0.13	+0.13		= +0.13
Seasonal deviation	−1.12	1.24	4.01	−4.12		

(all figures to 2 decimal places)

Thus expense claims are at their highest in quarter 3 (on average £4,010 above the trend) and at their lowest in quarter 4 (on average £4,120 below the trend).

Question 3.2

(a) Since a Laspeyre index is required, the number of units used in 1986 is not relevant to the calculation.

$$\Sigma p_n q_o = (2.42 \times 10,400) + (22 \times 500) + (2.40 \times 5,000) = £18,168$$
$$\Sigma p_o q_o = (1.90 \times 10,400) + (18 \times 500) + (2 \times 5,000) = £38,760$$

If 1982 = 100, the cost index for 1986 is

$$\frac{48,168}{38,760} \times 100 = 124.3$$

(b) To compare the two series of index numbers the base year of one of the series needs to be changed. Changing the base year of the price index to 1982 involves dividing every number in the series by 86 and multiplying by 100. The results are shown below.

Year	1982	1983	1984	1985	1986
Price index	100	104.7	116.3	130.2	133
Cost index	100	102	109	116	124.3

This shows that the company's production costs have been rising at less than the general rate of inflation.

Question 4.1

(a) (i) p(both take 5 weeks to complete) $= 0.7 \times 0.6 = 0.42$

(ii) p(both activities completed in 4 weeks or less)
= p(A takes 4 weeks) $\times p$(B takes 3 or 4 weeks)
= $0.1 \times 0.4 = 0.04$

(iii) p(both activities have the same completion time)
= p(both take 4 weeks) + p(both take 5 weeks)
+ $(0.1 \times 0.2) + (0.7 \times 0.6) = 0.44$

(iv) p(A takes longer than B)
= p(A takes 6 weeks)
+ p(A takes 5 weeks and B takes 3 or 4 weeks)
+ p(A takes 4 weeks and B takes 3 weeks)
= $0.2 + (0.7 \times 0.4) + (0.1 \times 0.2)$
= 0.5

(b) Expected completion time of A = $(4 \times 0.1) + (5 \times 0.7) + (6 \times 0.2)$
= 5.1 weeks

Expected completion time of B = $(3 \times 0.2) + (4 \times 0.2) + (5 \times 0.6)$
= 4.4 weeks

The expected completion time is the mean completion time which would result if the activity was repeated a large number of times.

Question 4.2

The decision table for the problem is shown below (a decision tree could have been used instead)

Investment chosen	Change in interest rates		(Returns)
	Fall	Rise or remain same	
A	£5,000	£5,000	
B	£3,000	£7,000	
C	£1,000	£8,000	

Expected return on A = £5,000
Expected return on B = 0.4(£3,000) + 0.6(£7,000) = £5,400
Expected return on C = 0.4(£1,000) + 0.6(£8,000) = £5,200

Thus B yields the highest expected return. However, note the closeness of the expected values: it would take only a slight change in the estimated probability for A or C to give the highest expected payoff. Also, the attitude to risk of the decision maker has not been considered.

Question 5.1

(a) Two damaged units would cost £80 to replace while three would cost £120. Therefore, for replacement costs to exceed £100, three or more units must be damaged so we require

$$p(3 \text{ or more units damaged out of } 120) = 1 - [p(0) + p(1) + p(2)]$$

Using the Poisson approximation to the binomial distribution $n = 120$ and $p = 0.02$.
Therefore mean no. of damaged units $= m = 0.02 \times 120 = 2.4$.

$$p(r \text{ successes}) = \frac{e^{-m} m^r}{r!} \text{ (or using tables e.g. ICMA table 9)}$$

$$p(0) = \frac{e^{-2.4}(2.4)^0}{0!} = 0.0907$$

$$p(1) = \frac{e^{-2.4}(2.4)^1}{1!} = 0.2177$$

$$p(2) = \frac{e^{-2.4}(2.4)^2}{2!} = 0.2613$$

$$\text{so } p(0) + p(1) + p(2) = 0.5697$$

Therefore $p(3 \text{ or more units damaged}) = 1 - 0.5697 = 0.4303$.

(b) If the Poisson approximation to the binomial distribution is valid, this would imply that all units had both an equal and an independent chance of being damaged during delivery. The fact that one unit has been damaged has no effect on the probability of other units being damaged.

Question 6.1

(a) $\bar{x} = 10$ kg, $s = 2$ kg and $n = 30$. 95% confidence interval for the mean $= \bar{x} \pm 1.96\, s/\sqrt{n} = 10 \pm 1.96\,(2)/\sqrt{30} = 10 \pm 0.72$ kg i.e. 9.28 to 10.72 kg.
There is a 0.95 probability that this interval will contain the mean weight of all the bags in the warehouse.

(b) The inspector requires a confidence interval with a precision of ± 0.5 kg therefore

$$\frac{1.96\,(2)}{\sqrt{n}} = 0.5$$

(assuming that 2 kg is a reasonable estimate of the population standard deviation). Cross multiplying

$$\frac{1.96\,(2)}{0.5} = \sqrt{n}$$

therefore $n = 61.47$.

Thus sixty-two bags would need to be sampled to obtain the required precision (sixty-one bags would give a confidence interval which was slightly too wide).

Question 6.2

A test of a hypothesis concerning the population proportion is required here. π = the claimed proportion of buses which arrived late = 0.12; $n = 200$, p = sample proportion = 30/200 = 0.15.
(i) Set up hypotheses: $H_0 \pi = 0.12$; $H_1 \pi > 0.12$ (i.e. a one-tail test).
(ii) Level of significance: 5% (since the question does not state a level of significance you could have used another level).
(iii) Decision rule: reject H_0 if Z exceeds 1.65.

(iv) Calculations: $Z = \dfrac{p - \pi}{\sqrt{\dfrac{\pi(1-\pi)}{n}}} = \dfrac{0.15 - 0.12}{\sqrt{\dfrac{0.12(1-0.12)}{200}}}$

$$= \frac{0.03}{0.023} = 1.31$$

Therefore do not reject the H_0 since 1.31 does not exceed the critical value. The sample does not provide evidence that the bus company are exaggerating their claim.

Question 6.3

A chi-squared test can be used here.
(i) Set up hypotheses: H_0 There is *no* difference between the regions in the proportion of customers who experienced late delivery. H_1 The proportion of customers who experienced late delivery does differ between the regions.
(ii) Level of significance: 5% (again, other values would be permissible).
(iii) Decision rule: Since there are 2° of freedom, reject H_0 if χ^2 exceeds 6.0.

(iv) Calculations: The table below shows the observed values with the expected values in brackets.

	Sales region		
	North England	Scotland	Midlands
Not experienced late delivery	12 (10.8)	30 (27)	12 (16.2)
Experienced late delivery	8 (9.2)	20 (23)	18 (13.8)

$$\text{therefore } \chi^2 = \frac{(12-10.8)^2}{10.8} + \frac{(30-27)^2}{27} + \frac{(12-16.2)^2}{16.2}$$

$$+ \frac{(8-9.2)^2}{9.2} + \frac{(20-23)^2}{23} + \frac{(18-13.8)^2}{13.8}$$

$$= 3.38$$

so the H_0 is not rejected. The sample does not provide sufficient evidence for the company to conclude that the proportion of customers experiencing late delivery varies between the regions.

Question 7.1

(a) Profit = revenue − costs; therefore
monthly profit = $P = -2x^2 + 80x - 300 - 24x = -2x^2 + 56x - 300$.
(b) $\mathrm{d}P/\mathrm{d}x = -4x + 56 = 0$ at the turning point, therefore $x = 14$ at the turning point. $\mathrm{d}^2 P/\mathrm{d}x^2 = -4$, so that profits are maximized at 14 units of output.
(c) At the break-even point(s), monthly cost = monthly revenue, i.e.
$300 + 24x = -2x^2 + 80x$, therefore $2x^2 - 56x + 300 = 0$.
Using the formula for finding the roots of a quadratic

$$x = \frac{56 \pm \sqrt{(-56)^2 - 4(2)(300)}}{2(2)}$$

$$= \frac{56 \pm 27.1}{4} = \text{either } 7.23 \text{ or } 20.78$$

so that the company will break even at output levels of either 7.23 or 20.78 units per month.

Question 8.1

(a) (i) $A + B = \begin{bmatrix} 20+5 & 4+3 \\ 2+4 & 18+0 \end{bmatrix} = \begin{bmatrix} 25 & 7 \\ 6 & 18 \end{bmatrix}$

(ii) Since a 2×2 matrix is to be multiplied by a 2×2 matrix, another 2×2 matrix will result

Row 1 of A × column 1 of B

20	×	5	=	100
4	×	4	=	16

element in row 1, col 1 of new matrix = 116
116

Similarly, Row 1 of A × column 2 of B = element in row 1, col 2 of new matrix = 60
Row 2 of A × column 1 of B = element in row 2, col 1 of new matrix = 82
Row 2 of A × column 2 of B = element in row 2, col 2 of new matrix = 6

Therefore $A \times B = \begin{bmatrix} 116 & 60 \\ 82 & 6 \end{bmatrix}$

Question 8.2

The cash flows (in £s) are set out below (it is assumed that they occur at the year end).

Year	Inflow	Outflow	Net cash flow	Discounted cash flow
0	0	−30,000	0	−30,000
1	25,000	16,000	9,000	0.89 × 9.000 = 8,010
2	25,000	16,000	9,000	0.80 × 9.000 = 7,200
3	25,000	16,000	9,000	0.71 × 9.000 = 6,390
4	31,000	16,000	15,000	0.64 × 15,000 = 9,600

Net present value = £1,200

Thus the vehicle does seem worth purchasing. (Note that the CIMA present value tables were used here. The discounting could also have been carried out by using the formula

Question 8.3

The compound interest formula is $S = P(1+r)^n$. Therefore after three years the money will have accumulated to $2,000(1+0.08)^3 = £2,519.42$. This money will then earn interest at 10 per cent for the next two years so that at the end of year 5 it will be worth $2,519.42(1+0.1)^2 = £3,048.50$.

Question 9.1

(b) $r = \dfrac{n\Sigma xy - (\Sigma x)(\Sigma y)}{\sqrt{[n\Sigma x^2 - (\Sigma x)^2][n\Sigma y^2 - (\Sigma y)^2]}}$

$= \dfrac{8(211) - (103)(28.9)}{\sqrt{[8(1,785) - (103)^2][8(164.03) - (28.9)^2]}}$

$= \dfrac{-1,288.7}{\sqrt{(3,671)(477.03)}} = -0.9738$

Therefore, as might be expected, there is a strong negative correlation between the two variables: as temperature rises, heating costs decrease.

(c)   The coefficient of determination $= r^2 = (-0.9738)^2 = 0.9484$. This means that 94.84 per cent of the variation in heating costs between the weeks can be explained by variations on the mean temperature at midday.

(d)   For the regression line:

$b = \dfrac{n\Sigma xy - (\Sigma x)(\Sigma y)}{n\Sigma x^2 - (\Sigma x)^2} = \dfrac{8(211) - (103)(28.9)}{8(1,785) - (103)^2} = \dfrac{-1,288.7}{3,671} = -0.35$

and

$a = \dfrac{\Sigma y}{n} - \dfrac{b\Sigma x}{n} = \dfrac{28.9}{8} - (-0.35)\dfrac{103}{8} = 8.12$

so the regression equation is: $y = 8.12 - 0.35x$.

This shows that, if the mean midday temperature for a week is 0°C, then heating costs of £812 would be expected. Each one-degree increase in the mean midday temperature will decrease costs by an expected £35.

(e)   If $x = 16$, $y = 8.12 - 0.35(16) = 2.52$, i.e. heating costs of £252 would be anticipated. Note that the line must be interpreted with care. It predicts negative heating costs at a temperature of 25°C.

Question 9.2

Spearman's rank correlation coefficient can be used here. This requires that *both* sets of data are ranked as shown below:

Package	User-friendliness rank	Cost rank (1 = most expensive)	d	d^2
A	5	4	1	1
B	2	5	-3	9
C	1	1	0	0
D	4	2	2	4
E	3	3	0	0
F	6	7	-1	1
G	7	6	1	1
	$n = 7$		$\Sigma d = 0$	$\Sigma d^2\ 16$

therefore $r' = 1 - \dfrac{6\Sigma d^2}{n(n^2 - 1)} = 1 - \dfrac{6(16)}{7(7^2 - 1)} = 0.714$

This suggests a fairly strong association between the cost and user-friendliness of the packages. Generally, the more expensive packages are the most user-friendly.

Question 10.1

The linear programming model is formulated as follows.
(i) Identify the decision variables. Let x = the number of kg of ingredient X used in the food per day, let y = the number of kg of ingredient Y used in the food per day.
(ii) Formulate the objective function: Minimize $C = 1.7x + 0.8y$ (where C = the daily cost of the ingredients).
(iii) Identify and formulate the constraints. Protein constraint: The number of kg of X and Y used must between them contain at least 8 units of protein. Each kg of X contains 2 units of protein while each kg of Y contains 1 unit. Therefore the protein constraint is: $2x + y \geq 8$. Similarly, the iron constraint is: $6x + y \geq 12$ and the vitamin D constraint is: $x + 3y \geq 9$. Also, the non-negativity constraints are: $x \geq 0$ and $y \geq 0$.

Since there are two decision variables, the problem can be solved graphically (see Figure A2). Note that the feasible area lies 'above' the constraints (because they are 'greater than' constraints) and, as defined by the model, it is unbounded to the 'north' and 'east' (though of course there must be some limit to the number of kg of the two ingredients which can be fed to the animal in a day!).

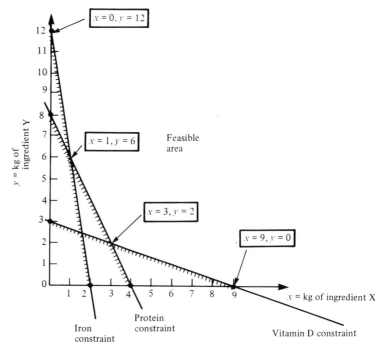

Figure A2

The combinations of ingredients which might lead to the minimum cost are given below, together with their costs.

x	y	Cost = 1.7x + 0.8y
0	12	£9.60
1	6	£6.50
3	2	£6.70
9	0	£15.30

Thus if 1 kg of ingredient X and 6 kg of Y are fed to the animal each day, costs will be minimized.

Question 10.2

(a) The model is: Maximize $C = 20x_1 + 25x_2 + 40x_3$, subject to: $4x_1 + 5x_2 + 6x_3 \leq 228$ and $x_1 + x_2 + 2x_3 \leq 80$. $x_1 \geq 0$, $x_2 \geq 0$, $x_3 \geq 0$ (where x_1, x_2 and x_3 equal the number of Standard, De-luxe and Cordon-bleu cookers produced per week and C = contribution per week).

(b) (i) Produce no Standard or De-luxe models, but thirty-eight Cordon-bleu models (this reflects the relatively high contribution of Cordon-bleu models).

(ii) There are no spare person-hours available, but there are four type E components spare (this can be verified: thirty-eight Cordon-bleu cookers, each using two type E components, will require seventy-six components per week out of the eighty which are available).

(iii) The optimal contribution is £1,520 (again, this can be verified: thirty-eight Cordon-bleu cookers at £40 per unit will generate £1,520).

(iv) If an extra person-hour could be made available it would increase total weekly contribution by £6.67. Since there are already spare type E components, their shadow price is £0.

(v) The contribution of each Standard cooker must increase by £6.67 to make it worth producing. The contribution of the De-luxe model must increase by £8.33 per unit.

Question 11.1

(a) The Economic order quantity $= \sqrt{\dfrac{2C_o D}{C_h}}$

Here $C_o = £50$, $D = 9 \times 50 = 450$ units, $C_h = 15\%$ of £30 $= £4.50$, therefore

$$\text{EOQ} = \sqrt{\frac{2(50)450}{4.5}} = 100 \text{ units}$$

so the current ordering policy is not optimal.

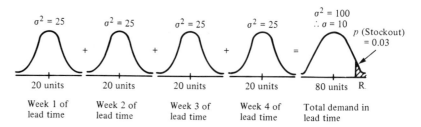

Figure A3

At EOQ: annual holding costs + annual ordering costs = average stock $\times C_h$ + no. of orders $\times C_0$ = $100/2 \times 4.5 + 450/100 \times 50$ = £450.

With current policy: annual holding costs + annual ordering costs = $200/2 \times 4.5 + 450/200 \times 50$ = £562.50. Therefore an annual saving of £112.50 could be made by changing the re-order quantity to the EOQ. (At the EOQ there will be 4.5 orders per year. This can be interpreted as nine orders every two years.)

(b)   To find the probability of demand during the lead time, four normal distributions need to be summed, as shown in Figure A3. The resulting distribution has a mean of eighty units and a variance of 100. Therefore the standard deviation is $\sqrt{100}$ = 10 units.

If R = the re-order level which will ensure no more than a 0.03 probability of running out of stock then (from the tables) R has a Z value of about 1.88. This means that R is 1.88 standard deviations above the mean demand. Thus $R = 80 + 1.88(10) = 98.8$, i.e. a re-order level of ninety-nine units will achieve the objective (this implies a safety stock of nineteen units).

Question 11.2

Note that costs are incurred for the time the lorry spends in the system. For the current team: mean time a lorry spends in the system = $1/(\mu - \lambda)$ where $\lambda = 4$ per hour and $\mu = 60/10 = 6$ per hour (be careful here $\mu \neq 10$) therefore the mean time a lorry spends in system = $1/(6-4) = 1/2$ hour, therefore the mean cost per lorry = $1/2 \times £6 = £3$. But, on average, four lorries arrive per hour therefore the mean cost per hour = $4 \times £3 = £12$; and cost of employing 3 people per hour = $3 \times £2.50 = £7.50$. Total cost per hour = $12 + £7.50 = £19.50$.

If the team size is increased: $\lambda = 4$ per hour, $\mu = 60/8 = 7.5$ per hour, so the mean time a lorry spends in system = $1/(7.5-4) = 0.286$ hours. Therefore the mean cost per lorry = $0.286 \times £6 = £1.71$. But, on average, four lorries arrive per hour, therefore the mean cost per hour = $4 \times £1.71 = £6.86$, and cost of employing 4 people per hour = $4 \times £2.50 = £10.00$. Total cost per hour = $6.86 + 10 = £16.86$.

Therefore the team should be extended to four people since this will lead to an average saving of £2.64 per hour.

Question 12.1

(a) and (b)   For the network and activity times see Figure A4 (note that a dummy end node has been added with zero duration because there is more than one end activity). The estimated project duration is twenty-six weeks.

(c)   (i) If eight weeks elapse before B can start (rather than the six which were originally estimated) then this will have no effect on the project's duration because B has three weeks' total float. This float would, of course, be reduced to one week by the delayed start.

(ii) If F takes nine weeks to complete the project's duration will be extended by two weeks to twenty-eight weeks. This is because F is a critical activity and any delay in its completion will delay the entire project.

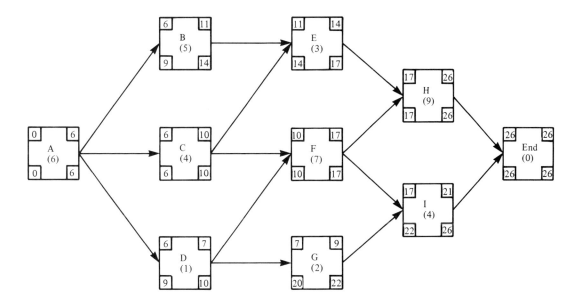

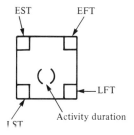

Figure A4

Part Four
Examination Technique

Poor performance in examinations is sometimes caused by faulty examination technique. The notes set out below give a few simple suggestions as to how you should prepare for the examination, and how you should tackle the paper itself on the day.

Before the examination

1 Keep an eye on the journal of your Association or Institute. This will contain a section where announcements are made about examinations (e.g. the type of calculator which can be used, changes to the rubric of papers, syllabus clarifications etc.).
2 Check that you know the correct time and place of the examination.
3 Make sure that you have obtained all the equipment that you will need for the examination. Your list will probably include a couple of pens and pencils, a ruler, a rubber, a watch and a calculator.

Your calculator should at least have a square root key (for the standard deviation) and, ideally, an x^y key, or its equivalent (for calculating binomial probabilities, compound interest etc.). For some examinations (e.g. the ACCA(2.6) Quantitative Analysis paper) it would also be advisable to have a calculator with an e^x key (for Poisson probabilities) and a 'log' and 'antilog' key (for non-linear regression). Some accountancy bodies now allow programmable calculators to be used in their examinations.

The examination

1 Read carefully the instructions that are given on the front page of the question paper. Some papers are divided into sections and you will lose marks if you answer the wrong number of questions from each section. In particular, look out for compulsory questions. Special instructions may also be given for multiple choice questions (see below).
2 When you are clear about the number of questions to be answered, read through *all* the questions on the paper before choosing which questions to answer. Then, as a confidence-booster, select the easiest question and prepare to answer this first. This approach has a number of advantages. In particular, it will enable your mind to work subconsciously on the more difficult questions while you are answering the easier ones.
3 Before starting to write, re-read the first question you have selected and

think about how the answer should be approached. Make sure that you understand clearly what the question is asking for. It is all too easy to answer the question which you think is being asked rather than the one which has actually been set.

While reading through long questions, it is a good idea to highlight, or note, the main points as you come to them. For example, in a linear programming question, where the production level of two products has to be determined, you may find it useful to draw up a rough table recording the contribution of the products, the resources they use and the constraints.

If the question requires you to write an essay, examiners will award marks for well-structured, clear and logical answers. It is therefore worth spending some time planning your answer. Jot down in rough the main issues which you intend to deal with and then arrange them into a logical order. This will enable you to develop your arguments lucidly and coherently.

When you start writing make sure that all your points are relevant to the question. Examiners are not impressed by waffle and you will be wasting time by writing it.

4   A surprising number of candidates waste valuable time by copying the questions, word for word, into their answer books. You will earn no extra marks by doing this.

5   If you are using graphs in your answer, make sure that all the axes are labelled and that all units (e.g. £, tonnes etc.) are clearly stated. In tables, all rows and columns should have clear headings.

6   In questions requiring calculations, show all your workings. Calculators are available which will work out measures like the standard deviation automatically. However, if no workings are displayed, and your answer is wrong, the examiner will not know whether the mistake was owing to a silly error or a lack of understanding of the subject. He would not, therefore, be able to award any marks for your answer.

7   Many parts of questions are designed to test your ability to interpret and convey the results of quantitative analysis and a substantial number of marks can be gained by answering these questions well. Nevertheless, there is sometimes a tendency on quantitative methods papers for candidates to treat the written sections of questions less seriously than sections requiring calculations. While you should answer the written parts of questions concisely, one- or two-word answers are unlikely to be adequate. For example, if you are asked to interpret a scattergraph, an answer like 'the scattergraph indicates negative correlation' will earn few marks. You should use the answer to demonstrate to the examiner that you know what negative correlation is, and what its implications are in the context of the question.

It is worth noting here that questions which ask you to discuss the limitations, or assumptions, of a particular technique are very common indeed.

8   Your allocation of time between questions should reflect the marks which can be gained on the questions. (Some examinations tell you how much time to spend on particular sections.) Ignoring this rule is a very common fault. The first few marks on questions are the easiest to obtain. It is therefore not worth spending more than your allotted time trying to 'squeeze' a few extra marks from a question when you could obtain easier marks by moving on to other questions. You should, of course, attempt all the questions which are required.

9 If you find that you are running out of time, list the main points or outline the methods which you would have used in your answer.

10 Always stay until the end of the examination. If you finish early, use the remaining time to check your answers for errors and omissions.

Multiple choice questions

Some of the accountancy bodies have introduced (or are thinking of introducing) sections of multiple choice questions into their examinations. To reflect this, multiple choice questions have been included in the test paper at the end of this book.

A typical example of a multiple choice question might be

> The variance of the numbers: 7, 7, 7, 7, 7 is
> *A*: 0 *B*: 7 *C*: 49 *D*: 245

Candidates are given a special answer sheet and, for each question, they are asked to circle the letter which represents the answer of their choice.

The main implications of these questions for students are

1 The questions are an objective test in that they eliminate any bias on the part of the person marking the paper. All that is required from the student is a tick or a circle around the particular option. This means that factors like poor handwriting do not disadvantage the candidate.

2 The questions are normally compulsory and they enable the examiner to test a wide variety of topics quickly. It is therefore unwise to concentrate your revision on just a few areas of the syllabus.

3 The distractors (these are the options given in the question which are incorrect) are answers which would be obtained if a common mistake was made. They are all designed to appear to be plausible. You should therefore be careful to check your answer before selecting an option.

Part Five
Test Paper

Test paper

This paper is in two sections. CACA students taking syllabus 1.5 (Business Mathematics and Information Technology) should answer only the questions in Section I while CACA students taking syllabus 2.6 (Quantitative Analysis) should answer Section II. You should answer the test under examination conditions and allow yourself forty minutes for each section. Answers to the questions are given on page 172.

Section I

1 For each of the following multiple choice questions circle the letter which indicates the answer of your choice.

(i) The median of 3, 1, 0, 2, 5, 13, 4 is

   A: 2 B: 3 C: 3.5 D: 7

(ii) If a price index (with 1980 = 100) has the value 120 in January 1985 and 150 in January 1986, prices have increased between January 1985 and January 1986 by

   A: 25% B: 30% C: 50% D: 150%

(iii) If $y = 3x^3 + 2x - 20$, at $x = 2$ dy/dx is equal to

   A: 8 B: 20 C: 38 D: 40

(iv) £200 is invested today and earns 10 per cent compound interest per annum. At the end of two years the money will have accumulated to

   A: £210 B: £220 C: £240 D: £242

2 The figures below refer to the weekly wages of a group of employees.

Wages (£)	No. of employees
80 to under 100	23
100 to under 120	31
120 to under 140	10
140 to under 160	5
160 to under 180	1
	70

(a) Calculate the mean wage.
(b) Calculate the standard deviation of the wages.
(c) Draw on ogive and estimate the interquartile range.

3   A company is selling a piece of property and the purchaser is offering three alternative methods of payment

Method 1: A single payment now of £18,000.
Method 2: Three payments: £12,000 now, £6,000 in a year's time and £2,000 in two years' time.
Method 3: An annuity of £4,000, payable at the end of each of the next six years.

Assuming that a discount rate of 12 per cent is applicable, calculate the present value of the three methods of payment. Which method is likely to be preferred by the company?

4   The total costs (C) per week of producing a product depend upon the level of output (x) as shown by the equation below

$$C = 300 + 10x$$

The total revenue received per week (R) is also dependent upon the level of output as shown by the following equation

$$R = 90x - 3x^2$$

The level of output never exceeds twenty-five units in a week. Determine (i) the break-even point(s) (to the nearest whole number); (ii) the level of output where revenue is maximized.

Section II

5   The size of an advertisement, placed in a weekly newspaper in five different weeks, is shown below, together with the number of readers who replied each week to the advertisement.

Size of advertisement (column inches) (x)	2	4	5	7	4
No. of replies received (y)	90	160	230	310	130

(a)   Derive the linear regression equation which could be used to predict the number of replies which would be received for an advertisement of a given size.
(b)   Forecast the number of replies which would be received if a six-column inch advertisement was placed in the newspaper.

6   (a)   If a company's current stock control policy continues, it is estimated that there is a probability of 0.1 that it will run out of stock in a given week. Assuming that the binomial distribution is applicable to the problem determine the probability that the company will run out of stock in: (i) two of the next five weeks, (ii) less than two out of the next five weeks.

(b)   Eighty bills are selected at random from the large number which were sent out by a company in a given month, and the time before payment is received is recorded. For the sampled bills, the mean time before payment was 60.2 days, with a standard deviation of 12 days. Estimate the mean time before payment of all the bills which were sent out in the month at the 95 per cent level of confidence.

(c) What would be the effect on the confidence interval if the sample size was quadrupled to 320 bills?

7   A project consists of seven activities, details of which are given below

Activity	Preceding activity	Estimated duration (days)
A	–	4
B	A	8
C	A	7
D	B and C	3
E	C	9
F	D and E	2
G	E and F	1

(a) Find the critical path, and hence determine the project's duration.
(b) Calculate the total float of each activity.
(c) What would be the effect of (i) activity D taking nine days, instead of the estimated three? (ii) seven days elapsing before activity B can start? (Consider these two modifications separately.)

8   A company is planning its production for next month and has to determine how many units of products A and B it should produce in order to maximize contribution. Each unit of A produced sells for £160 and each unit of B sells for £200. Both products are made from the same raw material which cost £4 per kilogram. Each unit of A produced requires 30 kg of the material and each unit of B requires 44 kg. However, only 6,000 kg of the material are available per month.

The products also require a special packing material. Only 300 rolls of this can be obtained per month at a cost of £14 per roll. Each unit of A produced requires 2 rolls while each unit of B requires 1.5 rolls. It is considered necessary to produce at least 20 units of A and no more than 100 units of B in the month. Formulate the problem as a linear programming model. (Do not solve the problem.)

Answers to test paper

Section I

1 (i) *B* (ii) *A* (iii) *C* (iv) *D*

2 (a) Mean wage = £110

(b) Standard deviation = £18.82

(c) Q1 = £95 (approx.), Q3 = £119 (approx.) therefore IQR = £24 (approx.)

3

Method	Present Value
1	£18,000
2	£18,952 (answers may vary slightly because of
3	£16,445 rounding)

Therefore method 2 appears to be the one which is most likely to be preferred.

4 (i) Break-even levels of output are five and twenty-two units to nearest whole number, (ii) Revenue is maximized at an output level of fifteen units.

Section II

5 (a) $y = -20 + 46.36x$

(b) about 258 replies

6 (a) (i) 0.0729 (ii) 0.9185

(b) 60.2 ± 2.63 days i.e. 57.57 to 62.83 days.

(c) The width of the interval would be halved.

7 (a) Critical path is A C E F G. The project's duration is twenty-three days.

(b)

Activity	A	B	C	D	E	F	G
Total float (days)	0	5	0	5	0	0	0

(c) (i) The project completion would be delayed by one day to twenty-four days.

(ii) There would be no effect since B has five days' float and is only being delayed by three days.

8 If a = no. of units of A produced in the month; b = no. of units of B produced in the month and C = contribution the model is: Maximize $C = 12a + 3b$ subject to: $30a + 44b \leq 6,000$; $2a + 1.5b \leq 300$; $a \geq 20$ and $b \leq 100$. Also $b \geq 0$.

Appendix 1 Area under the normal curve

This table gives the area under the normal curve between the mean and a point z standard deviations above the mean. The corresponding area for deviations below the mean can be found by symmetry.

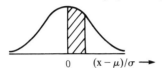

$0 \quad (x - \mu)/\sigma \longrightarrow$

$\frac{(x - \mu)}{\sigma}$	0.00	0.01	0.02	0.03	0.04	0.05	0.06	0.07	0.08	0.09
0.0	.0000	.0040	.0080	.0120	.0159	.0199	.0239	.0279	.0319	.0359
0.1	.0398	.0438	.0478	.0517	.0557	.0596	.0636	.0675	.0714	.0753
0.2	.0793	.0832	.0871	.0910	.0948	.0987	.1026	.1064	.1103	.1141
0.3	.1179	.1217	.1255	.1293	.1331	.1368	.1406	.1443	.1480	.1517
0.4	.1554	.1591	.1628	.1664	.1700	.1736	.1772	.1808	.1844	.1879
0.5	.1915	.1950	.1985	.2019	.2054	.2088	.2123	.2157	.2190	.2224
0.6	.2257	.2291	.2324	.2357	.2389	.2422	.2454	.2486	.2518	.2549
0.7	.2580	.2611	.2642	.2673	.2704	.2734	.2764	.2794	.2823	.2852
0.8	.2881	.2910	.2939	.2967	.2995	.3023	.3051	.3078	.3106	.3133
0.9	.3159	.3186	.3212	.3238	.3264	.3289	.3315	.3340	.3365	.3389
1.0	.3413	.3438	.3461	.3485	.3508	.3531	.3554	.3577	.3599	.3621
1.1	.3643	.3665	.3686	.3708	.3729	.3749	.3770	.3790	.3810	.3830
1.2	.3849	.3869	.3888	.3907	.3925	.3944	.3962	.3980	.3997	.4015
1.3	.4032	.4049	.4066	.4082	.4099	.4115	.4131	.4147	.4162	.4177
1.4	.4192	.4207	.4222	.4236	.4251	.4265	.4279	.4292	.4306	.4319
1.5	.4332	.4345	.4357	.4370	.4382	.4394	.4406	.4418	.4430	.4441
1.6	.4452	.4463	.4474	.4485	.4495	.4505	.4515	.4525	.4535	.4545
1.7	.4554	.4564	.4573	.4582	.4591	.4599	.4608	.4616	.4625	.4633
1.8	.4641	.4649	.4656	.4664	.4671	.4678	.4686	.4693	.4699	.4706
1.9	.4713	.4719	.4726	.4732	.4738	.4744	.4750	.4756	.4762	.4767
2.0	.4772	.4778	.4783	.4788	.4793	.4798	.4803	.4808	.4812	.4817
2.1	.4821	.4826	.4830	.4834	.4838	.4842	.4846	.4850	.4854	.4857
2.2	.4861	.4865	.4868	.4871	.4875	.4878	.4881	.4884	.4887	.4890
2.3	.4893	.4896	.4898	.4901	.4904	.4906	.4909	.4911	.4913	.4916
2.4	.4918	.4920	.4922	.4925	.4927	.4929	.4931	.4932	.4934	.4936
2.5	.4938	.4940	.4941	.4943	.4945	.4946	.4948	.4949	.4951	.4952
2.6	.4953	.4955	.4956	.4957	.4959	.4960	.4961	.4962	.4963	.4964
2.7	.4965	.4966	.4967	.4968	.4969	.4970	.4971	.4972	.4973	.4974
2.8	.4974	.4975	.4976	.4977	.4977	.4978	.4979	.4980	.4980	.4981
2.9	.4981	.4982	.4983	.4983	.4984	.4984	.4985	.4985	.4986	.4986
3,0	.49865	.4987	.4987	.4988	.4988	.4989	.4989	.4989	.4990	.4990
3.1	.49903	.4991	.4991	.4991	.4992	.4992	.4992	.4992	.4993	.4993
3.2	.49931	.4993	.4994	.4994	.4994	.4994	.4994	.4995	.4995	.4995
3.3	.49952	.4995	.4995	.4996	.4996	.4996	.4996	.4996	.4996	.4997
3.4	.49966	.4997	.4997	.4997	.4997	.4997	.4997	.4997	.4997	.4998
3.5	.49977									

Index